D0213523

MARGARET CAVENDISH

Philosophical Letters

Abridged

EARLY MODERN & MODERN WOMEN PHILOSOPHERS

MARGARET CAVENDISH

Philosophical Letters

Abridged

Edited by

Deborah Boyle

Hackett Publishing Company, Inc.
Indianapolis/Cambridge

24 23 22 21   1 2 3 4 5 6 7

For further information, please address
        Hackett Publishing Company, Inc.
        P.O. Box 44937
        Indianapolis, Indiana 46244-0937

        www.hackettpublishing.com

Cover design adapted by E. L. Wilson
Interior design by Elana Rosenthal
Composition by Aptara, Inc.

Library of Congress Control Number: 2020950103

ISBN-13: 978-1-62466-974-3 (cloth)
ISBN-13: 978-1-62466-973-6 (pbk.)

The paper used in this publication meets the minimum requirements of American National Standard for Information Sciences—Permanence of Paper for Printed Library Materials, ANSI Z39.48–1984.

∞

CONTENTS

Introduction   vii

    Cavendish's Life and Works   xi

    Cavendish's Philosophical System   xv

    Cavendish's Opponents   xxvii

A Note on the Text   xlix

Margaret Cavendish: Her Life, Her Times   li

Suggestions for Further Reading   lv

Philosophical Letters   1

    A Preface to the Reader   3

    Section 1   7

    Section 2   81

    Section 3   129

    Section 4   175

Index   225

INTRODUCTION

In a time when few women published their writings, even anonymously, Margaret Cavendish (1623–1673) published more than a dozen books under her own name. In an era dominated by theories that treated matter as inert, she argued that matter is intrinsically self-moving and perceptive; in a culture that viewed the natural world as created for human use, she maintained that humans were no more special than other creatures, and that to think otherwise was hubris. She was the first woman to be allowed to visit the Royal Society of London, where she saw the experiments and scientific apparatus of the "new science" of the seventeenth century. Although she described herself as bashful around others,[1] she was bold in print.

For all of this, Cavendish gained a certain notoriety in her lifetime; a 1667 entry of Samuel Pepys's famous diary reported that "the whole story of this lady is a romance."[2] Cavendish hoped for lasting fame. In one of her early works, *Worlds Olio* (1655), she wrote that she hoped fortune would put her book "in fame's high tower," where it would, "like a cannon bullet . . . make so loud a report, that all the world shall hear it."[3] Instead, her writings—including not just philosophy but also poetry, plays, stories, and orations—fell mostly into oblivion for more than three hundred years. To the extent that she was known, it was for her own autobiography and her biography of her husband, William Cavendish, Duke of Newcastle.[4] Those who did know of Cavendish's other works were often less than complimentary; while conceding that Cavendish's writings were

---

1. Margaret Cavendish, "A True Relation of My Birth, Breeding, and Life," in *Natures Pictures* (1656), 373.

2. Samuel Pepys, entry of April 11, 1667. See Robert Latham, ed., *The Shorter Pepys* (London: Bell and Hyman, 1985), 754.

3. Margaret Cavendish, "A Dedication to Fortune," in *Worlds Olio* (1655), sig. A1v.

4. The biography was *The Life of the Thrice Noble, High and Puissant Prince William Cavendishe, Duke, Marquess and Earl of Newcastle* (1667). A Latin translation (1668) and English reprint (1675) were also published. Mark Antony Lower's 1872 republication of the first edition, with Cavendish's autobiography, led to additional reprints in the late nineteenth and early twentieth centuries.

animated by a "vein of authentic fire," Virginia Woolf nonetheless characterized her as "crack-brained and bird-witted."[5]

Only in the past few decades has Cavendish's work been subject to serious
scholarly consideration. Her poetry, plays, and 1666 work of science fiction, *The
Description of a New World, Called the Blazing World*, started to receive attention
in the 1980s, with the focus on her views about gender.[6] Cavendish's writings in
natural philosophy were next to be taken seriously, although again the focus was
on her views on gender and how those informed her natural philosophy.[7] This
is an important element of Cavendish's work, to be sure, but foregrounding her
views on gender meant that Cavendish's natural philosophy was not yet being
interpreted in its own terms; it meant reading her as a "woman philosopher" and
not just as a philosopher. This began to change in the 1990s, as historians of
philosophy started to see Cavendish's philosophy as an interesting philosophical
system in its own right.[8]

This system was first laid out in Cavendish's 1653 book, *Philosophicall Fancies*. She expanded this work into the 1655 *Philosophical and Physical Opinions*;
she then revised the 1655 book for a 1663 second edition, which was, in turn,
the basis for Cavendish's final treatise in natural philosophy, *Grounds of Natural
Philosophy* (1668). Along the way, she published two more works in natural philosophy, *Philosophical Letters* (1664) and *Observations upon Experimental Philosophy*
(1666), which was published with its companion piece, *Blazing World*. Some of
these volumes have now been published in modern editions, and selected letters
from *Philosophical Letters* have appeared in anthologies of her works. This edition
of *Philosophical Letters* aims to make this important text available to students,
teachers, and scholars in an accessible, abridged format.

---

5. Virginia Woolf, "The Duchess of Newcastle," in *The Common Reader: First Series* (New York:
Harcourt, Brace & World, 1925), 78.

6. For example, see Dolores Paloma, "Margaret Cavendish: Defining the Female Self," *Women's
Studies: An Interdisciplinary Journal* 7, no. 1/2 (1980): 55–66; and Sylvia Bowerbank, "The Spider's
Delight: Margaret Cavendish and the 'Female' Imagination," *English Literary Renaissance* 14, no. 3
(1984): 392–406.

7. For example, see Lisa Sarasohn, "A Science Turned Upside Down: Feminism and the Nat-
ural Philosophy of Margaret Cavendish," *Huntington Library Quarterly* 47, no. 4 (1984): 289–307;
Rebecca Merrens, "A Nature of 'Infinite Sense and Reason': Margaret Cavendish's Natural Phi-
losophy and the 'Noise' of a Feminized Nature," *Women's Studies* 25, no. 5 (1996): 421–38; and
John Rogers, "Margaret Cavendish and the Gendering of the Vitalist Utopia," in *The Matter of
Revolution* (Ithaca: Cornell University Press, 1996), 177–211.

8. For example, see Susan James, "The Philosophical Innovations of Margaret Cavendish," *British
Journal for the History of Philosophy* 7 (1999): 219–44.

Philosophical Letters is worth reading for several reasons. First and foremost, it is a valuable source for understanding Cavendish's own philosophical system. Cavendish herself emphasizes in the Preface that her purpose in examining the works of others is "to make my own opinions the more intelligible" (*PL* b1v/3).[9] Thus we find in *Philosophical Letters* detailed discussions of her views and arguments regarding self-moving matter, perception, the relation between God and nature, and other aspects of her natural philosophy and epistemology. Indeed, whereas in her other works of natural philosophy Cavendish often simply *states* her views, *Philosophical Letters* contains much more *defense* of those views.

But because Cavendish is directly engaging with the views of her contemporaries, the letters also show that she was not working in an intellectual vacuum; she was well versed in the work of Thomas Hobbes, René Descartes, Henry More, and others. While Cavendish notes that juxtaposing her views against others' makes her own views clearer, doing so helps us understand the views she is criticizing too. To read *Philosophical Letters* is to get a clearer picture of the shape of philosophy in the seventeenth century.

Finally, *Philosophical Letters* is fascinating for its genre.[10] As Gary Schneider has observed, letters in this era were often not intended only for sender and recipient, but were "collective social forms designed, understood, and expected to circulate within designated epistolary circles."[11] Printed volumes of letters by foreign or classical authors such as Jean-Louis Guez de Balzac, Cicero, and Seneca were widely read in early modern England, and, Schneider points out, "after 1645 collections of personal letters and letters of state by Englishmen were

---

9. All parenthetical references are to Margaret Cavendish, *Philosophical Letters: Or, Modest Reflections Upon Some Opinions in Natural Philosophy, Maintained by Several Famous and Learned Authors of this Age, Expressed by Way of Letters* (1664), followed by page numbers for this volume. Unnumbered pages in Cavendish's Preface are identified by the lower-case letters and numbers used as printers' marks for binding the book, with "r" to indicate the right-hand page and "v" to indicate the left.

10. Diana Barnes examines *Philosophical Letters* in light of humanist epistolary theory in "Familiar Epistolary Philosophy: Margaret Cavendish's *Philosophical Letters*," *Parergon* 26, no. 2 (2009): 39–64. David Cunning argues that the structure of *Philosophical Letters* reflects Cavendish's views about women's agency in a patriarchal setting; see "Cavendish, *Philosophical Letters*, and the Plenum" (forthcoming in *Cavendish: An Interdisciplinary Perspective*, ed. Lisa Walters and Brandie Siegfried [Cambridge: Cambridge University Press]).

11. Gary Schneider, *Culture of Epistolarity: Vernacular Letters and Letter Writing in Early Modern England, 1500–1700* (Newark: University of Delaware Press, 2005), 22.

published with increasing frequency."[12] But Schneider's use of "Englishmen" is telling: *women's* correspondence was typically not published.[13]

Some published correspondence of the early modern period focused on philosophy; for example, Robert Boyle presented his book of accounts of experiments with the air pump in the form of a letter to his nephew Lord Dungarvan,[14] and *Usefulness of Experimental Philosophy* (1663) was written as a letter to another nephew, Richard Jones.[15] Henry Oldenburg, the secretary of the newly founded Royal Society of London, established the journal *Philosophical Transactions* in 1665, using it to publish scientific papers written in the form of letters addressed to him and submitted from natural philosophers throughout Europe.[16] Much important philosophical work was carried on in correspondence by women writers; among the best known letters today are those from Princess Elisabeth of Bohemia to Descartes, but she was by no means the only seventeenth-century woman writing letters about philosophy.[17] Again, however, these letters among women were not *published* in that era.[18] So Cavendish's choice, as a woman, to publish letters on philosophy was striking.

It is also striking that Cavendish's female correspondent was (as far as we know) fictional. While the genre of the epistolary novel only took off in the eighteenth century, publications of letters to or from imaginary characters were not

---

12. Schneider, *Culture of Epistolarity*, 49.

13. James Daybell notes that sometimes husbands had their wives' letters printed posthumously; see "Gendered Archival Practices and the Future Lives of Letters," in *Cultures of Correspondence in Early Modern Britain 1550–1642*, ed. James Daybell and Andrew Gordon, (Philadelphia: University of Pennsylvania Press, 2016), 210–36, at 213.

14. Robert Boyle, *New Experiments Physico-Mechanicall, Touching the Spring of the Air, and Its Effects* (1660).

15. See Lawrence M. Principe, "Virtuous Romance and Romantic Virtuoso: The Shaping of Robert Boyle's Literary Style," *Journal of the History of Ideas* 56, no. 3 (1995): 377–97, at 394.

16. See Marie Boas Hall, "The Royal Society's Role in the Diffusion of Information in the Seventeenth Century," *Notes and Records of the Royal Society of London* 29, no. 2 (1975): 173–92.

17. See Carol Pal, *Republic of Women: Rethinking the Republic of Letters in the Seventeenth Century* (Cambridge: Cambridge University Press, 2012).

18. Princess Elisabeth's side of the correspondence with Descartes was only published in 1879, after being discovered in a castle in the Netherlands. See Lisa Shapiro, ed. and trans., *The Correspondence between Princess Elisabeth of Bohemia and René Descartes* (Chicago: University of Chicago Press, 2007). Anne Conway only corresponded with Henry More after insisting that their letters not be made public; see Sarah Hutton, *Anne Conway: A Woman Philosopher* (Cambridge: Cambridge University Press, 2004). For further examples of unpublished correspondence by seventeenth-century women philosophers, see Jacqueline Broad, ed., *Women Philosophers of Seventeenth-Century England: Selected Correspondence* (New York: Oxford University Press, 2020).

uncommon in the seventeenth; examples are Nicholas Breton's *A Post With a Packet of Mad Letters* (1602), Madeleine de Scudéry's *Amorous Letters from Various Contemporary Authors* (1641), and Cavendish's own *CCXI Sociable Letters* (1664). Occasionally these publications covered philosophical topics. To take Robert Boyle as an example again, his 1659 book *Some Motives and Incentives to the Love of God* (often referred to as *Seraphic Love*) was published as a letter to a fictional correspondent named Lindamor.[19]

Philosophical Letters may well be unique for the way it combines preexisting elements in one work: they are letters exclusively on philosophical topics, penned from a real woman philosopher to an imaginary female correspondent, and published so that the public could read them. As Jacqueline Broad has noted, Cavendish also adapted the genre of correspondence to suit her own purposes in *Philosophical Letters*:

> Cavendish completely disregards the typical conventions of letter writing. Her letters are undated and without location; she never exchanges any personal details or topical news; and the quantity of letters is unusually large. . . . The correspondence serves as convenient rhetorical device.[20]

The voice in *Philosophical Letters* is entirely Cavendish's own, and she is entirely in control of the direction that the correspondence takes.

Cavendish's Life and Works

Margaret Cavendish was born in 1623 to a landowning family in Colchester, England.[21] Cavendish's father, Thomas Lucas, died when she was two, so her mother took on the task of raising the eight children—of whom Cavendish was the youngest—and running the estate. As Cavendish reports in her autobiography,

---

19. For discussion, see Principe, "Virtuous Romance and Romantic Virtuoso," 382–83, and Lawrence M. Principe, "Style and Thought of the Early Boyle: Discovery of the 1648 Manuscript of Seraphic Love," *Isis* 85, no. 2 (1994): 247–60.

20. Jacqueline Broad, "Cavendish, van Helmont, and the Mad Raging Womb," in *The New Science and Women's Literary Discourse: Prefiguring Frankenstein*, ed. Judy A. Hayden (New York: Palgrave Macmillan, 2011), 47–63, at 54.

21. Many of the following details are drawn from Katie Whitaker's superb biography, *Mad Madge: The Extraordinary Life of Margaret Cavendish, Duchess of Newcastle, the First Woman to Live by Her Pen* (New York: Basic Books, 2002).

she received the kind of education considered suitable for a young lady, learning to read, write, sing, dance, and embroider; she received no formal education in subjects such as Latin or Greek, history, philosophy, mathematics, or logic.[22] She loved to go on long walks that she spent in "musing, considering, contemplating," and she loved to read and write.[23] If we can take her first-person claims in CCXI *Sociable Letters* as autobiographical, she wrote sixteen little books as a child; she calls them her "baby books," both because they were made of folded-up sheets of paper and because "it was in my baby-years I writ them."[24]

Cavendish's life was irrevocably changed by the English Civil War. She and her family were members of the Anglican Church and supporters of King Charles I. Indeed, Margaret's brother John was a tax collector for the king, and it was, in part, the king's collection of taxes without parliamentary approval that led to public opposition to the royal regime.[25] Moreover, Essex, where the Lucases lived, was highly supportive of Parliament and the Puritans.[26] In 1642, with war imminent, local residents poured into the Lucas estate, and, as Cavendish tells it, "plundered" her family "of all their goods, plate, jewels, money, corn, cattle, and the like, cut down their woods, pulled down their houses, and sequestered them from their lands and livings."[27] Her brother John was imprisoned in the Tower of London for a month; Cavendish's other two brothers were already away, fighting for the king. Cavendish herself went to London, where two of her married sisters lived.[28]

Soon thereafter, at the age of twenty, Cavendish made a decision that surprised her family: she joined the court of Queen Henrietta Maria as a maid of honor. This meant moving to Oxford, where the king was conducting business and the queen was ensconced in Merton College. The duties of maids of honor were minimal, for they were primarily expected to wait in an anteroom until needed by the queen, and to stand at attention in the formal dining room while

---

22. Cavendish, "A True Relation," 370–71.

23. Cavendish, "A True Relation," 386–87.

24. Margaret Cavendish, CCXI *Sociable Letters* (1664), letter 131. For a modern edition of this work, see Margaret Cavendish, *Sociable Letters*, ed. James Fitzmaurice (Peterborough, Ontario: Broadview Press, 2004).

25. Whitaker, *Mad Madge*, 34–35.

26. Whitaker, *Mad Madge*, 34.

27. Cavendish, "A True Relation," 377. For a fuller account, see Whitaker, *Mad Madge*, 38–41.

28. Whitaker, *Mad Madge*, 42.

the queen ate.[29] Cavendish found that she disliked the gossipy environment, but her mother insisted that she continue her with commitment.[30] The country was in upheaval, and the war was going poorly for the Royalists; the Scottish had joined the Parliamentarian side, and in 1644 they invaded England. With the Parliamentary army advancing on Oxford, the queen and her entourage made a harrowing escape to Paris.

Cavendish did not continue long in her position as maid of honor after the move to Paris, for that year she met William Cavendish, Marquess of Newcastle (1593–1676). An extremely wealthy widower with five grown children, William was thirty years older than Margaret. He had served as an army commander for the king, but had been badly defeated in battle at Marston Moor in July 1644. After the defeat, he fled into exile in France, a move that cost him his reputation.[31] In Paris, however, he made himself useful to the queen, and there he noticed and began courting Margaret. They married in late 1645.

William Cavendish was a well-educated man who, before the war, had read widely, written plays and poetry, and maintained friendships with prominent philosophers, poets, and playwrights, such as Thomas Hobbes and Ben Jonson. He was interested in natural philosophy and established a laboratory at one of his homes, Bolsover.[32] These interests continued after he settled in Paris, where many of his friends had also fled, so Margaret met many of the luminaries of the era. Visitors to their home included philosophers René Descartes, Pierre Gassendi, Marin Mersenne, Kenelm Digby, Bishop John Bramhall, and Hobbes,[33] as well as poets Edmund Waller and John Davenant.[34] William continued to pursue his interests in natural philosophy, acquiring multiple telescopes; Margaret had access to these as well as her own microscope.[35]

In 1648, Margaret and William moved to the Netherlands. With the Parliamentarians in control in England, King Charles I was being held captive at the Isle of Wight and would be executed in 1649. Now deemed a traitor to the country, William was unable to return to England, so he and Margaret stayed

---

29. Whitaker, *Mad Madge*, 48 and 50.

30. Cavendish, "A True Relation," 374.

31. Whitaker, *Mad Madge*, 67–68.

32. On William Cavendish, see Whitaker, *Mad Madge*, 63–69.

33. In 1645, Hobbes and Bramhall debated free will and necessity at William's home, an exchange that later appeared in print in a series of pamphlets. See Vere Chappell, ed., *Hobbes and Bramhall on Liberty and Necessity* (Cambridge: Cambridge University Press, 1999).

34. Whitaker, *Mad Madge*, 90–94.

35. Whitaker, *Mad Madge*, 98–99.

in the Netherlands, first in Rotterdam and then in Antwerp. Margaret did make one trip back to England, accompanied by William's brother Charles; as William's wife, she was entitled to petition Parliament for access to his money, so she traveled to England in 1651, remaining there for eighteen months. During this time she published her first two books, *Poems, and Fancies* (1653) and *Philosophicall Fancies* (1653). The appeal to Parliament was unsuccessful; Margaret returned to Antwerp empty-handed.[36] The couple remained in Antwerp as Oliver Cromwell came to power in England, and Cavendish continued to write. A collection of very brief essays, observations, and allegories that she had worked on in 1651 came out as *Worlds Olio* in 1655, as did a revised version of *Philosophicall Fancies*, which she called *Philosophical and Physical Opinions*. *Natures Pictures* (1656) was a collection of stories and included her autobiography.

In 1658 Oliver Cromwell died, and in 1660 Charles II was installed as king. At that point the couple were finally able to return to England. William's estates, Welbeck and Bolsover, had been managed in his absence by his adult children, but many of his possessions had had to be sold, and the forests and hunting parks were in ruins. While William focused on returning his estates to their former glory, Cavendish embarked on a reading program. As biographer Katie Whitaker has noted, a bill for books Cavendish purchased from London in 1664 would have covered the cost of between one and two hundred books. These likely included some of the books that Cavendish discusses in *Philosophical Letters*.

Cavendish wrote and published prolifically between 1662 and 1671. In 1662 she published a volume of plays, as well as *Orations of Divers Sorts*, an assortment of speeches presented by fictional characters in various contexts, including a fascinating set of seven speeches by women to each other about gender roles.[37] The year 1663 saw a second edition of *Philosophical and Physical Opinions*, a book to which Cavendish frequently refers in *Philosophical Letters*, where she calls it "my book of philosophy." In 1664, besides *Philosophical Letters*, Cavendish also published *CCXI Sociable Letters* and a second edition of *Poems, and Fancies*. *Observations upon Experimental Philosophy* and *Description of a New World, Called the Blazing World* were published as companion pieces in 1666; a biography of her husband came out in 1667; *Grounds of Natural Philosophy*, second editions of *Observations* and *Blazing World*, and a new volume of plays came out in 1668; and

---

36. Whitaker, *Mad Madge*, 130–34.

37. Margaret Cavendish, *Orations of Divers Sorts* (1662), part 11. A modern edition of *Orations* appears in Susan James, ed., *Margaret Cavendish: Political Writings* (Cambridge: Cambridge University Press, 2003).

second editions of *Natures Pictures* and *Worlds Olio* were published in 1671. Cavendish died in 1673 at the age of fifty and was buried in Westminster Abbey.

Cavendish's Philosophical System

Cavendish's primary goal in *Philosophical Letters* is to contrast her own philosophical views with those of others, on the assumption that her correspondent is already familiar with the theory of matter and corresponding epistemology expounded in *Philosophical and Physical Opinions*. Nonetheless, the key features of her system can be gleaned without reading either her prior or later works of natural philosophy.

What Cavendish seeks to explain is nothing less than nature itself. What is nature made of? Is everything material, or are there immaterial entities like souls? What is God's relationship to the natural world? How do processes in nature occur—that is, how can we best explain natural phenomena, ranging from human and animal action and perception, to diseases and cures, magnetism and light, snowfall and earthquakes? And how sure can we be about the accuracy of those explanations—can we have genuine knowledge about nature? What can we know, if anything, about God?

Cavendish's Account of Nature

Cavendish's account of nature can be understood from three perspectives. We can consider what she says about (1) nature as a whole, which she calls "infinite matter" or "only matter"; (2) the mixed stuff of which nature as a whole is composed; and (3) the individual entities into which that stuff is divided. I consider each of these perspectives in turn.

(1) Nature as a Whole

Cavendish maintains that nature is made of one single unified stuff: matter. Nature is a plenum, with no empty spaces or vacuum (PL 7/11). This means that matter is continuous; although some portions of matter may move differently from other portions, no part of matter can actually be separated from other parts. Matter is infinitely extended as well as eternal. It is also all that exists in nature; Cavendish denies the existence of immaterial substances in nature.

The infinitude of nature is an important theme in *Philosophical Letters*; Cavendish takes up the issue near the beginning and returns to the topic three more times, including in the penultimate letter. In these letters, she is not so much *arguing for* the thesis of nature's infinitude as *defending it against* objections, particularly against objections that her view is inconsistent with Christian doctrine that nature is God's creation.[38] But her responses are also designed to show that belief in an omnipotent God *supports* her thesis. As she points out in *Observations upon Experimental Philosophy*, if God is eternal, then "his actions are so, too, the chief of which is the production or creation of Nature."[39] In *Philosophical Letters*, this argument is expressed in her claim that "God, being infinite, cannot work finitely," and thus that matter must be infinite in size and duration (PL 458/199).

On the topic of immaterial substances, Cavendish's views can be found throughout *Philosophical Letters*, but especially in her responses to Henry More. One argument is based on the presupposition that matter is inherently in motion (a claim for which she offers other arguments; see below). She diagnoses belief in an immaterial substance as stemming from some philosophers' mistaken separation of the concept of motion from the concept of matter, leading them to think that motion can exist without matter; this matter-less motion is their concept of an immaterial substance (PL 11–12/13–14). But she also argues that an immaterial substance in nature is actually inconceivable (PL 69/48). Those who think they have a concept of the immaterial are mistaken; to use the word "immaterial" about something in nature is really just nonsense (PL 320–21/154).[40] This does leave open the possibility of immaterial divine souls that are *not* part of nature, but Cavendish insists that since these are supernatural, beyond nature, they are not really conceivable or understandable (PL 186–87/109–10).

---

38. On Cavendish's attempt to reconcile the theses that matter is eternal and that God created matter from nothing, see Deborah Boyle, "Margaret Cavendish on the Eternity of Created Matter," in *Early Modern Women on Metaphysics*, ed. Emily Thomas (Cambridge: Cambridge University Press, 2018), 111–30.

39. Margaret Cavendish, "Further Observations upon Experimental Philosophy," in *Observations upon Experimental Philosophy*, 2nd ed. (1668), chapter 14. For a modern edition, see Eileen O'Neill, ed., *Observations upon Experimental Philosophy* (Cambridge: Cambridge University Press, 2001).

40. Compare this with Hobbes, who writes in part 3, chapter 34 of *Leviathan* (1651) that *"substance incorporeal* are words which, when they are joined together, destroy one another" (Thomas Hobbes, *Leviathan*, ed. Edwin Curley [Indianapolis: Hackett Publishing Co., 1994], 262).

(2) Nature as a Mixture

A second perspective on Cavendish's conception of nature focuses on the qualities of the material stuff that makes up the "infinite matter." While Cavendish holds that nature is entirely material (and to that extent she is a monist), she also holds that this matter is composed of three "degrees." Two of these degrees are "animate"; that is, they are self-moving, able to initiate motion in themselves without being compelled by anything external to them. One of these is "rational" matter, the other "sensitive" matter. The third degree of matter is inanimate, and only in motion because of its mixture with the animate degrees. Indeed, it is a key part of Cavendish's view that inanimate and animate matter are completely blended together, constituting a commixture (*PL* 98–99/63, 444/190, 461/200).[41] Since all of nature is composed of this matter, all of nature is both rational and perceptive (*PL* 18–19/18).

Cavendish saw the relationship among the three degrees of matter as hierarchical.[42] Rational matter is like an architect or designer, planning and commanding what is to be done; the sensitive parts of matter are like laborers, the "architectonical, and working parts" that are supposed to carry out the commands of the rational parts (*PL* b2v/5, 150/89, 416–17/176); the inanimate matter is the material used by the sensitive parts to perform their motions.[43] According to Cavendish, then, actions in the natural world occur through a command issued by a bit of rational matter, which is then (ordinarily) carried out by a bit of sensitive matter, using the inanimate matter as its materials. But, again, these three "degrees" of matter are always completely intermixed.

(3) Nature as a Collection of Creatures

Nature can also be considered as a collection of finite parts, or what Cavendish calls "creatures" (*PL* 433/184).[44] Although Cavendish believed nature to be one continuous whole, she also recognized that it is composed of a great many (she

---

41. Jonathan Shaheen discusses Cavendish's doctrine of blended matter in "Part of Nature and Division in Cavendish's Materialism," *Synthese* 196 (2019): 3551–75.

42. In Cavendish's day, "degree" could also mean "social rank."

43. Cavendish explicitly characterizes inanimate matter this way in later texts. See Margaret Cavendish, *Grounds of Natural Philosophy* (1668), part 1, chapter 5; and part 2, chapter 4. For a modern edition, see Anne Thell, ed., *Grounds of Natural Philosophy* (Peterborough, Ontario: Broadview Press, 2020).

44. Alison Peterman characterizes these as "effective" parts of matter (that is, as effects of motion), in contrast to the three degrees of matter, which are "constitutive" parts, in "Cavendish on Motion and Mereology," *Journal of the History of Philosophy* 57, no. 3 (2019): 471–99.

would say infinite) individual entities such as rocks, trees, horses, and humans, as well as artifacts made by humans, such as chairs and ships. This also includes the smaller parts of those things (leaves and bark, for example) as well as higher-level collections of which those things are themselves parts (such as a forest, or, for humans, a family or a church congregation).

Cavendish's view is that particular finite items or creatures are individuated by how their motions differ, for matter is in constant motion, with an infinite variety of motions, both internal and external (PL 100/64). Motion, for Cavendish, encompasses far more than just what we might call "local motion," that is, change of location. Her views about how to classify the varieties of motion changed over time; in the first edition of *Philosophical and Physical Opinions*, she says there are four "common motions"—attractive, retentive, digestive, and expulsive[45]—while in the second edition she says there are six, adding "contraction" and "dilatation" to the earlier four.[46] In her later works—*Philosophical Letters, Observations*, and *Grounds*—she gives up trying to reduce all kinds of motion to four or six fundamental forms. However, she does still distinguish between the external motions of a thing and its internal motions. Roughly, external motions are those that cause changes in location or other perceptible alterations in a thing; for example, "dilating" and "contracting" motions make a thing change its color (PL 124/77) or its shape (PL 120/74). Internal motions, imperceptible to human sense, make an object the sort of thing it is and allow it to persist as that thing over time (PL 166/99); for example, it is a "retentive" motion that allows gold to remain gold even when melted, despite its change in shape and temperature (PL 409/172).

Individual creatures are differentiated from the matter around them by the differences in their motions, but individuals are also members of types, or what Cavendish calls "forms and figures" (PL 144/85). These "forms and figures" are differentiated by their typical or "proper" internal and external motions (PL 447/193, 539/221); so, for example, the typical internal and external motions of oak trees differ from those of palm trees, water, squirrels, or humans. Cavendish emphasizes that since there is an infinite number of varieties of motion, there is an infinite variety of things in the cosmos (PL 6/11).

Cavendish's idea, then, is that any given portion of matter has (a) motions that resemble those of certain other portions of matter, making it a member of a

---

45. Margaret Cavendish, *Philosophical and Physical Opinions* (1655), chapter 24.

46. Margaret Cavendish, *Philosophical and Physical Opinions* (1663), part 1, chapter 9. Cavendish's account of what these six motions do can be found in part 1, chapter 18.

certain type of thing, and (b) motions that are different enough from the motions of the matter around it to clearly differentiate it as being a distinct individual thing. The same portion of matter can also change its motions, thereby changing the "figure" that the motions produce; thus things can, over time, change into other things, a process that Cavendish elsewhere calls "transmutation."[47] When a creature dies, for example, the motion of the matter making up the creature's body changes, resulting in the decay of the body and its transformation into some other kind of entity (PL 218/123).

Cavendish thinks that her theory of matter as self-moving is more plausible than the kind of account endorsed by mechanists like Descartes and Hobbes, in which matter is naturally inert. On a mechanistic view, motion is local motion, and an object is only able to move when it is or has been the passive recipient of motion from some *other* object that is in motion. And because Cavendish rejects mechanistic accounts of motion, she also rejects mechanistic accounts of causal interaction between bodies. As Cavendish read Hobbes and Descartes, when one object A causes a change in another object B, object A transfers some of its motion to object B (PL 77–78/52).[48] But according to Cavendish, matter is inherently moving, so motion could *never* leave one body and be transferred to another, unless matter is also transferred, a process she calls "translation" (PL 420/179). Instead, Cavendish thinks causation is *occasional* causation: a body A serves as the occasional cause of a change in another body when the second body, B, responding to some action by A, produces the appropriate change in itself. She gives an example of a watchmaker setting a watch in motion:

> A watchmaker does not give the watch its motion, but he is only the occasion that the watch moves after that manner, for the motion of the watch is the watch's own motion, inherent in those parts ever since that matter was [. . .]. Wherefore one body may occasion another body to move so or so, but not give it any motion, but every body (though occasioned by another to move in such a way) moves by its own natural motion. (PL 100/64)

This account should be distinguished from the occasionalism that Nicolas Malebranche (1638–1715) and other later Cartesian philosophers endorsed, where *God* is the cause of effects in the natural world. As we shall see, Cavendish

---

47. See Cavendish, *Philosophical and Physical Opinions* (1663), part 4, chapters 26–28.

48. However, Cavendish may not have interpreted Hobbes and Descartes correctly; see fn81.

did not think God could play a role in the natural world.[49] How does one part of matter serve as an occasion for another to act, then? Cavendish maintains that matter can perceive and communicate information from one part to another (PL 181/107).[50] When one part of the natural world moves, the parts around it perceive that motion, and (typically) respond appropriately.

When the parts of nature move in their typical, normal ways, Cavendish calls such motions "regular." Of course, events in the physical world are not entirely uniform or predictable; sometimes surprising or unusual things happen. Cavendish characterizes these as "irregularities" (PL 359/166). For example, the human body is usually healthy; illness occurs, but it is not the usual state, and so Cavendish characterizes disease as an irregularity (PL 343/160). Cavendish insists that irregularities are natural (PL 359–60/166); indeed she sometimes even suggests that irregularities do not really exist and that "irregular" is just a label humans use when we perceive something unusual (PL 359–60/166 and 538–39/221). In other passages, she suggests that irregularities are genuine errors and disorders in nature (PL 29/25, 138/82, 152/90) that occur when a creature moves in a way not suitable to its kind (PL 183/107). How to resolve these tensions has been the subject of some debate in the secondary literature.[51]

Cavendish's Arguments for Self-Moving Matter

In positing that all matter is self-moving and perceptive, Cavendish was going against the grain of natural philosophy in her day. Descartes, Hobbes, Robert

---

49. Karen Detlefsen, "Reason and Freedom: Margaret Cavendish on the Order and Disorder of Nature," *Archiv für Geschichte der Philosophie* 89, no. 2 (2007): 157–91, at 166. See also Eileen O'Neill, "Margaret Cavendish, Stoic Antecedent Causes, and Early Modern Occasional Causes," *Revue Philosophique* 138, no. 3 (2013): 311–26.

50. Cavendish scholars disagree regarding whether Cavendish thinks the parts of nature require libertarian free will in order for occasional causation to operate. For libertarian interpretations, see Detlefsen, "Reason and Freedom," and Deborah Boyle, "Freedom and Necessity in the Work of Margaret Cavendish," in *Women and Liberty 1600–1800*, ed. Jacqueline Broad and Karen Detlefsen (Oxford: Oxford University Press, 2017), 141–62. For an interpretation of Cavendish as a compatibilist, see David Cunning, *Cavendish* (London: Routledge, 2016).

51. For interpretations arguing that Cavendish thinks there are no genuine disorders or irregularities, see Lisa Walters, *Margaret Cavendish: Gender, Science and Politics* (Cambridge: Cambridge University Press, 2014); Cunning, *Cavendish*; and Michael Bennett McNulty, "Margaret Cavendish on the Order and Infinitude of Nature," *History of Philosophy Quarterly* 35, no. 3 (2018): 219–39. For interpretations that read Cavendish as holding that there are genuine disorders in nature, see Detlefsen, "Reason and Freedom;" and Deborah Boyle, *The Well-Ordered Universe: The Philosophy of Margaret Cavendish* (New York: Oxford University Press, 2018).

Boyle, and others insisted that matter is inherently inert; for a physical thing to move required impact from some other entity, either another physical body or (for the occasionalists) from God. As Boyle put it in his 1674 defense of what he called his "mechanical" natural philosophy, its two "grand" principles were "matter, and motion; for matter alone, unless it is moved is wholly inactive."[52]

Cavendish uses the argumentative strategy of inference to the best explanation to defend her account of self-moving matter. When faced with two or more competing explanations for some phenomenon, inference to the best explanation involves arguing that one of the candidate explanations does a better job at meeting criteria such as simplicity and broad scope; the conclusion, then, is that that explanation is more likely to be true than the others.[53] Cavendish argues that explanations appealing to self-moving matter are better at explaining various observable features of the natural world than are mechanistic explanations.[54] When Cavendish considers the explanatory strength of some theory that others have proposed, she judges it less likely to be true than another if it either fails to predict some natural phenomenon that we experience, or does predict some phenomenon that we do not experience. She appeals to both large-scale and small-scale phenomena to make her case. Thus she says we see that processes in nature are generally predictable and orderly (*PL* 152/91), taking this as a reason to think that the parts of nature know what they are doing: "if none but man had reason, and none but animals sense, the world could not be so exact and so well in order as it is" (*PL* 44/34). That is, if it were true that only humans have reason (a view maintained by Descartes), we would not expect to see the order and predictability in the world that we do in fact observe; Cavendish takes it that her view, that reason pervades nature, does a better job of explaining order, and thus that her view is more likely to be true.

Focusing on more particular processes, Cavendish maintains that mechanistic explanations of human sense-perception in terms of matter in motion are implausible. Hobbes, for example, explained sensory perception in terms of the

---

52. Robert Boyle, "About the Excellency and Grounds of the Mechanical Natural Philosophy," in *The Excellency of Theology, Compared with Natural Theology* (1674), section 2. A modern edition appears in M. A. Stewart, ed., *Selected Philosophical Papers of Robert Boyle* (Indianapolis: Hackett Publishing Co., 1991).

53. For more on inference to the best explanation (sometimes called "abduction"), see Igor Douven, "Abduction," *Stanford Encyclopedia of Philosophy* online, ed. Edward N. Zalta (March 9, 2011; revised April 28, 2017), https://plato.stanford.edu/archives/sum2017/entries/abduction/.

54. Jacqueline Broad discusses another case where Cavendish uses inference to the best explanation in "Margaret Cavendish and Joseph Glanvill: Science, Religion, and Witchcraft," *Studies in History and Philosophy of Science Part A* 38, no. 3 (2007): 493–505.

pressure of external objects on the matter in a perceiver's sensory organs. As he puts it in *Leviathan*,

> The cause of sense, is the external body, or object, which presseth the organ proper to each sense, either immediately, as in the taste and touch, or mediately, as in seeing, hearing, and smelling; which pressure, by the mediation of nerves and other strings and membranes of the body, continued inwards to the brain and heart, causeth there a resistance, or counter-pressure, or endeavour of the heart, to deliver itself; which endeavor, because *outward*, seemeth to be some matter without.[55]

Cavendish's alternative account of perception agrees with Hobbes's account in one respect: sense-perceptions are constituted by motions. But Cavendish maintains that these motions are not caused by *pressure*. One of her arguments is that if sense-perception occurred by pressure, this would "cause such dents and holes [in the sense-organs and brain] as to make them sore and patched in a short time" (PL 22/20). Ears subject to loud music would "grow sore and bruised with so many strokes" (PL 72/49) and eyes seeing by the sun's light would be put in "as much pain as fire does, when it sticks its points into our skin or flesh" (PL 63/45). Injuries to ears and eyes do sometimes happen, of course, but Cavendish observes that even a pleasant sound would cause damage if it continued too long (PL 72/49). Since we do not suffer the kinds of injuries that the mechanistic account of sense-perception would lead us to expect, that is a strike against it.[56]

Cavendish's Epistemology

Cavendish's account of causation between bodies requires not just that matter can move itself, but that it is perceptive and knowing, for it is through perception of what occurs around it that any particular bit of matter knows how it should respond. If the motion of a ball is, as Cavendish maintains, the ball's own motion in response to the motion of the hand holding it, then the ball needs to know both what the hand has done and what it should do in response. How, then, does Cavendish propose to explain perception in terms of self-moving matter?

---

55. Hobbes, *Leviathan*, 6 (part 1, chapter 1); emphasis in the original.

56. On Cavendish's criticisms of the pressure model of perception, see Marcus Adams, "Visual Perception as Patterning: Cavendish against Hobbes on Sensation," *History of Philosophy Quarterly* 33, no. 3 (2016): 193–214; and Deborah Boyle, "'Informed by Sense and Reason': Margaret Cavendish's Theorizing about Perception," in *The Senses and the History of Philosophy*, ed. Brian Glenney and José Filipe Silva (New York: Routledge, 2019), 231–48.

In her 1666 *Observations*, Cavendish provides a more detailed epistemology in which all parts of matter have both perception and self-knowledge, but this distinction is not present in *Philosophical Letters*, where she explains only perception. She distinguishes between sensitive perception and rational perception; sensitive matter is capable of the former, rational matter of the latter. Since every portion of matter contains a blend of both sensitive and rational matter, every part possesses a "double perception" (PL 19/18, 115/73). By this, Cavendish means that every part of nature is *capable* of a double perception, not that every act of perception involves a double perception, for, as we shall see, sensitive and rational perception can come apart.

We have already seen that Cavendish rejected mechanistic accounts of perception in terms of the impact or pressure of external objects on the sense organs and nervous system. Her alternative theory, at least to explain *human* perception, appeals to both occasional causation and what she calls "patterning." To pattern something out is to copy or imitate it, which, for Cavendish, means copying its motions. Cavendish appeals to this process to explain not just perception but also phenomena such as the impression of an object in snow (PL 104–5/67), the reflection of an object in a mirror (PL 81/54), and echoes (PL 81/54).

As Cavendish explains human sense-perception, when some item is present to the human senses, the sensitive matter in the sense-organ of the perceiver patterns out or "figures" the motions of the object being perceived (PL 174–75/103). She describes how the sensitive matter in the eye responds to the presence of an external object such as a piece of embroidered silk:

> As for example, there is presented to sight a piece of embroidery, wherein is silk, silver, and gold upon satin in several forms or figures, as several flowers, the sensitive motions straight by one and the same act pattern out all those several figures of flowers, as also the figures of silk, silver, gold, and satin, without any pressure of these objects or motions in the medium. (PL 68/48)

The eye's perception of the embroidered silk is constituted by the motions in the sensitive matter as it copies the motions of the matter that comprises the piece of silk: the very act of patterning by the sensitive matter is the act of sense-perception.[57]

In most cases of human sense-perception, Cavendish thinks sensitive and rational perception work together; this is "double perception." Suppose, as in

---

57. Kourken Michaelian, "Margaret Cavendish's Epistemology," *British Journal for the History of Philosophy* 17, no. 1 (2009): 31–53, at 40.

Cavendish's example, someone is regarding a piece of embroidered silk. As the sensitive matter in the eye copies the patterns of the silk, the rational matter in the eye might copy patterns, too—not the patterns of the silk, but the motions of the sensitive matter that patterned out the silk. When rational matter does this, then there is "rational perception" of the silk. Cavendish does not clearly explain what additional purpose is served by rational perception when sensitive perception has also occurred, but elsewhere she hints that it involves *conscious awareness* of the sensation. In *Observations*, she gives an example of someone who is so engrossed in reading that she does not feel someone pinching her.[58] In this example, the person *does* have a sensitive perception of the pinch, yet, since rational matter does not copy the sensitive matter's actions, the person does not consciously feel the pinch. Elsewhere, Cavendish implies that the purpose of rational perception is to *unify* sensory perceptions coming in through multiple sense-organs so that the object is perceived as one.[59]

Just as sensitive perceptions can occur without being copied by motions of rational matter, Cavendish says rational perception can occur even if rational matter is not copying motions of sensitive matter. Any spontaneous motions of rational and sensitive matter (in the sense organs and brain, anyway) that are not imitating something else are said by Cavendish to be moving "by rote." Suppose the rational matter "figures" itself in ways that resemble an external object, but without copying currently occurring motions of the sensitive matter; if the rational matter repeats motions that *had* previously occurred in the sensitive matter, the person may be said to be remembering (PL 190/111); if it moves in completely new ways, "by rote," the person is imagining or conceiving (PL 173/102). Cavendish also suggests that emotions involve the rational matter moving on its own (PL 170/101). In dreaming and in madness, sensitive matter moves on its own, by rote, without copying any external object (PL 28/24, 73–74/50).

Cavendish emphasizes that her account of sensitive and rational perception in terms of patterning is meant to explain only human perception. But this does not mean that only humans perceive and think. Because all parts of nature are composed of the same intermixture of animate and inanimate matter, *all* of nature is self-moving, thinking, and perceptive (PL 18–19/18). Thus every creature, whether animal, vegetable, or mineral, is capable of both sensitive and

---

58. Cavendish, *Observations* (1668), chapter 36, "Of the Different Perceptions of Sense and Reason."

59. See Michaelian, "Margaret Cavendish's Epistemology," 42, and James, "Philosophical Innovations," 232. Cavendish suggests this in *Grounds*, part 1, chapter 10.

rational perception. The type of perception belonging to each kind of creature will depend on what kind of creature it is; that is, the knowledge possessed by humans differs from that possessed by other creatures (PL 75/51). Clearly, plants and minerals do not have the sensory organs that humans and many non-human animals have, so their perception can only be of a radically different kind than human perception.

Cavendish does not claim to know *how* nonhuman animals, plants, or inorganic matter perceive. Indeed, even regarding human perception as patterning, she does not claim that her account can be known with certainty to be true. She never claims to offer incontrovertible proofs of the tenets of her natural philosophy; she claims, rather, that her explanations are "probable" (see, for example, PL 11/13, 62/44, 81/54, 86/57, 133/79, 150/89). Cavendish thought absolute certainty is not achievable, for each of us is just a part of nature, limited in our "sense and reason" (PL 246/135). Thus natural philosophy can only be a search for the most probable explanations of natural phenomena; the more probable opinions are reasonable to hold, while the less probable ones should be rejected (PL 245/134).[60]

Cavendish's Philosophy of Religion

Cavendish thinks we can gain some knowledge, albeit only probable, of nature. What can we know about God? As we have seen, Cavendish thinks perception occurs through "patterning," where rational or sensitive matter moves in ways that copy, or imitate, the motions of the external object being perceived. But God is an infinite, incorporeal being (PL 8/11, 14/15), and so cannot be patterned by a finite, corporeal mind (PL 139/82). If something exists that a human mind cannot possibly pattern, then it seems human knowledge of that thing is impossible. Indeed, Cavendish insists that we cannot know God's essence (PL 140–41/83, 322/155).

However, Cavendish maintains that we can know that God *exists*, even if we cannot comprehend God's essence. She says that we all have "an idea, notion, conception, or thought of the existence of God" (PL 187/109); in fact, she insists that *all* parts of nature know that God exists, and even that all parts of nature love God (PL 138/81). But is Cavendish entitled to a distinction between ideas

---

60. Stephen Clucas explores Cavendish's probabilistic reasoning in "Variation, Irregularity and Probabilism: Margaret Cavendish and Natural Philosophy as Rhetoric," in *A Princely Brave Woman: Essays on Margaret Cavendish, Duchess of Newcastle*, ed. Stephen Clucas, 199–209 (Aldershot, Hampshire: Ashgate, 2003).

of the *essence* of God (which finite creatures cannot have) and ideas of God's *existence* (which she thinks finite creatures do have)? Doesn't having an idea of an entity's existence require having an idea that represents the entity itself? And given Cavendish's claims that finite corporeal matter cannot copy an infinite immaterial being, how can any part of matter have an idea that represents God?

Cavendish's reference to the "notion" of God suggests a solution. A "notion" is the result of rational matter moving "by rote," that is, forming certain figures that do not copy or pattern something else (*PL* 173/102). And Cavendish suggests that a figure formed by rational matter can represent a thing even *without* copying it; for example, she says that we can form thoughts of "infinite" because "the mind uses art, and makes such figures, which stand like to that" (*PL* 69/48). That is, even when we cannot perceive something like infinity, we can think about it using distinctive motions of the rational matter that *represent* that thing, although the motions do not *copy* or pattern anything.[61] So perhaps the idea of God's existence is a "notion" that represents God's existence *not* by actually copying God, but merely by "figuring" in some distinctive way that God has fixed or designated as representing himself.[62]

Cavendish's discussions of God in *Philosophical Letters* focus primarily on God's relationship to nature.[63] She maintains that while God is the creator of matter (*PL* 14/15, 16/16, 458–59/198), matter is nonetheless eternal—although she also suggests that "eternal" means something different when applied to God than when applied to matter. God is eternal insofar as God exists in "one fixed instant, without a flux or motion" (*PL* 455/196). Since matter is necessarily always in motion, it cannot be eternal in *that* sense. Nor can it be eternal in the sense of existing in a "fixed instant," for matter exists in time (*PL* 304/147), where "time" means "nothing else but the corporeal motions in nature" (*PL* 454/196).

---

61. Cavendish, *Philosophical and Physical Opinions* (1663), part 3, chapter 21.

62. Cavendish develops this suggestion further in her later works, where she characterizes knowledge of God as innate and "fixed"; see Cavendish, "An Argumental Discourse," in *Observations* (1668), sig. h3v. For further discussion of this suggestion, see Boyle, "Margaret Cavendish and the Eternity of Matter."

63. Lisa Sarasohn discusses Cavendish's views on other aspects of religion in "Fideism, Negative Theology, and Christianity in the Thought of Margaret Cavendish," in *God and Nature in the Thought of Margaret Cavendish*, ed. Brandie R. Siegfried and Lisa T. Sarasohn (Farnham, Surrey: Ashgate, 2014), 93–106.

To say that matter is eternal means, instead, that matter is "infinite in time or duration" (PL 459/199).[64]

But while Cavendish does have much to say about God in *Philosophical Letters*, she also emphasizes that she considers theology to be distinct from natural philosophy. She makes the point in her opening letter that she intends to "merely go upon the bare ground of natural philosophy and not mix divinity with it" (PL 3/8; see also PL 216/122, 316/152), and she sometimes takes aim at her philosophical opponents for not doing so themselves, as when she writes of Jan Baptist van Helmont that "he makes such a mixture of divinity and natural philosophy that all his philosophy is nothing but a mere hodge-podge, spoiling one with the other" (PL 248/136).[65] It is to these opponents that I turn next.

Cavendish's Opponents

Thomas Hobbes

After her opening letters on the infinitude of nature, Cavendish turns in section 1, letter 4 to the views of Thomas Hobbes (1588–1679).[66] She considers both *Leviathan* (1651) and *Elements of Philosophy, the First Section, Concerning Body* (1656), an English translation of Hobbes's 1655 *De Corpore*. Cavendish's analysis of *Leviathan* focuses on Hobbes's mechanistic, materialist natural philosophy,

---

64. David Cunning discusses Cavendish's claims about the eternity of matter in *Cavendish*, 163–64. For further discussion of how Cavendish conceived of God's relationship to matter, see Karen Detlefsen, "Margaret Cavendish on the Relation between God and World," *Philosophy Compass* 4, no. 3 (2009): 421–38; Sara Mendelson, "The God of Nature and the Nature of God," in *God and Nature in the Thought of Margaret Cavendish*, 27–42; and Boyle, *The Well-Ordered Universe*, 80–85.

65. On this theme in *Philosophical Letters*, see Stephen Clucas, "'A Double Perception in All Creatures': Margaret Cavendish's *Philosophical Letters* and Seventeenth-Century Natural Philosophy," in *God and Nature in the Thought of Margaret Cavendish*, 121–39.

66. For secondary literature relating Cavendish to Hobbes, see Sarah Hutton, "In Dialogue with Thomas Hobbes: Margaret Cavendish's Natural Philosophy," *Women's Writing* 4 (1997): 421–32; Lisa T. Sarasohn, "*Leviathan* and the Lady: Cavendish's Critique of Hobbes in the *Philosophical Letters*," in *Authorial Conquests: Essays on Genre in the Writings of Margaret Cavendish*, ed. Line Cottegnies and Nancy Weitz, 40–58 (Madison, New Jersey: Fairleigh Dickinson University Press, 2003); and Stewart Duncan, "Debating Materialism: Cavendish, Hobbes, and More," *History of Philosophy Quarterly* 29, no. 4 (October 2012): 391–409.

particularly as it applies to sensory perception, and other mental processes such as imagination, dreaming, understanding, and reasoning.

While it is Hobbes's political theory that primarily interests scholars today, Cavendish tells her reader she is not going to discuss this aspect of his thought. She cites three reasons for avoiding it: first, women are not involved in politics "unless an absolute queen" (PL 47/36), presumably a reference to Elizabeth I; second, even for a man, studying politics is pointless, unless he "were sure to be a favorite to an absolute prince"; and third, "it is but a deceiving profession, and requires more craft than wisdom" (PL 47/36). However, Cavendish's other writings show a persistent concern with politics, especially peace and war; this is not surprising, given the tumultuous period in which Cavendish was living.[67] As a good friend of William Cavendish, Hobbes was a regular visitor of the Cavendishes in Paris, so it seems likely that Margaret was familiar with his notions of the state of nature and the social contract as a way to escape it.[68] Indeed, a passage from her *Orations* bears a striking resemblance to Hobbes's famous description of the state of nature.[69] As Cavendish's orator puts it,

> If [there is] no safety [there is] no propriety, neither of goods, wives, children nor lives, and if there be no propriety there will be no husbandry and the lands will lie unmanured; also there will be neither trade nor traffic, all which will cause famine, war, and ruin, and such a confusion as the kingdom will be like a chaos, which the gods keep us from.[70]

---

67. See, for example, "The Inventory of Judgments Commonwealth" in *Worlds Olio*. Many of the speeches in *Orations* concern the best way to run a state, and in *The Description of a New World, Called the Blazing World*, the characters of the Empress and the Duchess discuss policies for good governance.

For secondary literature on Cavendish's political theory, see Ellayne Fowler, "Margaret Cavendish and the Ideal Commonwealth," *Utopian Studies* 7, no. 1 (1996): 38–48; Hilda Smith, "'A General War amongst the Men [but] None amongst the Women': Political Differences between Margaret and William Cavendish," *Politics and the Political Imagination in Later Stuart Britain*, ed. Howard Nenner (Rochester: University of Rochester Press, 1997), 143–60; Neil Ankers, "Paradigms and Politics: Hobbes and Cavendish Contrasted," in *A Princely Brave Woman*, 242–54; Mihoko Suzuki, *Subordinate Subjects: Gender, the Political Nation, and Literary Form in England, 1588–1688* (Aldershot, Hampshire: Ashgate, 2003), 182–202; and Deborah Boyle, "Fame, Virtue, and Government: Margaret Cavendish on Ethics and Politics," *Journal of the History of Ideas* 67, no. 2 (2006): 251–90.

68. Cavendish claimed to have barely ever spoken to Hobbes, but to have listened avidly to the philosophical discussions taking place around her (Whitaker, *Mad Madge*, 94 and 116).

69. Hobbes, *Leviathan*, 76 (book 1, chapter 13).

70. Cavendish, *Orations*, part 3, "An Oration against Liberty of Conscience."

And, again apparently echoing Hobbes, in a discussion of polity in *Worlds Olio*, Cavendish describes various contracts between the king and his subjects.

Nonetheless, Cavendish closes her discussion of *Leviathan* after addressing only the natural philosophical views from the first six chapters. At that point she turns to *Elements of Philosophy*; again, she skips topics that do not pertain to her own natural philosophical views, such as geometry—probably a wise decision, since even in his own day Hobbes's mathematical skills were derided.[71] In section 1, letters 16 and 17, Cavendish discusses Hobbes's accounts of bodies, accidents, place, and magnitude. She then returns to sense-perception, which Hobbes covers in much more detail in *Elements* than in *Leviathan*, with discussions of vision and light, hearing and echoes, and smell.

Even if Cavendish's responses to Hobbes are interesting mainly for how they illuminate her own views, a rough sketch of Hobbes's account of physical bodies may be helpful. Hobbes uses terms from medieval scholastic philosophy—often ultimately derived from Aristotle—that would have been familiar to seventeenth-century readers, but he does not use them in standard ways. To define "body," he appeals to "imaginary space," which scholastic philosophers from the late Middle Ages generally understood to mean an incorporeal void space outside the finite cosmos.[72] However, for Hobbes, "imaginary space" (or just "space") is what one imagines if one imagines all physical objects to be annihilated.[73] Strikingly, then, for Hobbes, space is merely a mental conception, not something real and absolute that exists independently of humans. Physical bodies, however, exist independently of us; Hobbes defines "body" as "that which having no dependence upon our thought is coincident or coextended with some part of space."[74] That is, we can think of a physical thing as, in a sense,

---

71. For an account of Hobbes's unsuccessful but persistent attempts to solve several important geometrical problems (and his unflagging conviction that his flawed approaches to these problems would work), see Douglas M. Jesseph, *Squaring the Circle: The War between Hobbes and Wallis* (Chicago: University of Chicago Press, 1999). Although acknowledging Hobbes's failure to solve these problems, Jesseph argues that Hobbes's account of geometry nonetheless has interesting features and is worth attention.

72. Some scholastic philosophers believed imaginary space actually existed. For detailed discussion, see Edward Grant, "Late Medieval Conceptions of Extracosmic ('Imaginary') Void Space," in *Much Ado About Nothing: Theories of Space and Vacuum from the Middle Ages to the Scientific Revolution* (Cambridge: Cambridge University Press, 1981), 116–47.

73. Thomas Hobbes, *Elements of Philosophy, the First Section, Concerning Body* (1656), part 2, chapter 7, section 2.

74. Hobbes, *Elements*, part 2, chapter 8, section 1.

overlapping (imaginary) space. The space exists only in our minds; the physical thing really exists outside us.

Matters get more confusing when Hobbes introduces the concept of "place." First, he notes that every body has a certain magnitude. The part of (imaginary) space with which a body's magnitude overlaps (or, in Hobbes's terminology, is "coincident") is the body's "place."[75] This means that when a body moves, its magnitude remains the same, but its place changes. In contrast, Cavendish holds that the place of a body in motion does *not* change; the place moves with the body. There is no spatial framework other than the plenum of self-moving matter, so "place" must be "an attribute which only belongs to a body" (PL 8/11), and there is no meaningful distinction to be drawn between "space," "place," and "magnitude" (PL 56/40).

Cavendish also takes aim at Hobbes's account of accidents. While the term "accident" has been interpreted in different ways throughout the history of philosophy, it typically means a *property* that a thing possesses, but does not possess *necessarily*. For example, Descartes took it to be an essential property of a piece of wax that it is extended, but an *accidental* property that it smells like flowers; if extension were somehow removed from the piece of wax, the wax itself would disappear, while the scent can be removed without changing the fact that the wax exists. Hobbes actually refers to this traditional understanding of "accident," noting that Aristotle held that "an accident is in its subject [. . .] so as that it may be away, the subject still remaining,"[76] but he then goes on to claim that some accidents are such that their removal *would* constitute a removal of the subject. A contemporary review of Hobbes's book expressed shock at this and other unorthodox claims,[77] but Cavendish is unperturbed by whether Hobbes's views contradict traditional scholastic views, for her interest, again, is with how well they fit with hers. For example, Hobbes claims that an accident such as the color red cannot be transferred from one object to another, but instead simply "perishes" in one object and is "generated" in another;[78] Cavendish notes her agreement that accidents cannot be transferred between objects, but denies that

---

75. Hobbes, *Elements*, part 2, chapter 8, section 5.

76. Hobbes, *Elements*, part 2, chapter 8, section 3.

77. Stephen Clucas, "An Early European Critic of Hobbes's *De Corpore*," *Hobbes Studies* 30 (2017): 4–27.

78. Hobbes, *Elements*, part 2, chapter 8, section 21.

accidents can be generated (*PL* 54–55/39).[79] Since she understands all properties of objects to be simply motions, and motions are necessarily connected with matter, for motion to cease to exist or to begin anew would mean that matter itself ceases or begins, a possibility Cavendish denies (*PL* 10–11/13). All changes in properties are simply changes in already-existing motions.

René Descartes

In section 1, letter 30, Cavendish begins discussing the work of the French philosopher René Descartes (1596–1650). Descartes's *Meditations on First Philosophy* (1641) is perhaps his most widely read work today, but there is no evidence that Cavendish had read the *Meditations*; her letters refer to Descartes's *Principles of Philosophy* (1644) and *Discourse on Method* (1637), along with the essays *Optics* and *Meteorology* that were published with the *Discourse*. She notes in her Preface that her knowledge of Descartes derived from a translation from the Latin that she had requested of some unnamed translator.[80]

Cavendish is mainly interested in Descartes's natural philosophy, primarily his accounts of motion and causal interaction between bodies. As we saw above, she reads Descartes as characterizing motion as a "mode" of material substance, so that when an object A moves another object B, the motion in A is transferred to B;[81] objecting that modes cannot be transferred between bodies (*PL* 98/62–63), Cavendish rejected this transfer account of motion.[82] Some later Cartesian philosophers embraced occasionalist theories in which God is the only cause of motion. Like Cavendish, they held that motion cannot be transferred; however, they retained the mechanistic idea that matter is inherently inert, concluding that God must cause motion because bodies cannot do so. Cavendish solved the

---

79. On Cavendish's account of colors, see Colin Chamberlain, "Color in a Material World: Margaret Cavendish against the Early Modern Mechanists," *Philosophical Review* 128, no. 3 (2019): 293–336.

80. While an English translation of the *Discourse* and associated essays had been published in 1649, *Principles of Philosophy* had not yet been translated into English by 1664. Cavendish's unknown translator probably relied on Latin versions of the *Principles* and the *Discourse, Optics,* and *Meteorology* that were published in Amsterdam in 1644.

81. See René Descartes, *Principles of Philosophy*, part 2, sections 25 and 40. This text is included in *The Philosophical Writings of Descartes*, vol. 1, trans. John Cottingham, Robert Stoothoff, and Dugald Murdoch (Cambridge: Cambridge University Press, 1985). Whether Descartes himself actually interpreted motion as the transfer of a mode is a vexed question in the secondary literature that need not detain us here.

82. Henry More had made a similar objection directly to Descartes; see More's letter of July 23, 1649, in *Oeuvres de Descartes*, vol. 5, ed. Charles Adam and Paul Tannery (Vrin: Paris, 1996), 382.

problem in a very different way, by jettisoning the idea that matter cannot be self-moving. Bodies can cause motion, but they cause their *own* motion.

Cavendish also takes on Descartes's theory of circular movement. According to Descartes's second law of motion, motion continues in a straight line unless some external cause interferes with it,[83] yet he also maintained that all motion is, ultimately, circular.[84] The latter claim depends on his view that nature is a plenum, without any vacuum; the motion of every part of matter is constrained by the matter around it. Thus, as Daniel Garber notes, despite Descartes's law that states that the tendency of all matter is to move in a straight line unless interfered with, "in his plenum, this condition of noninterference can *never* be met," and so all motion is circular.[85]

Cavendish considers Cartesian dualism only later, in letter 35, objecting to Descartes's claim that mind and body are separable. She responds by distinguishing between the "natural" soul and the "supernatural or divine" soul. Cavendish had in fact already conceded in her remarks on Hobbes that the "divine" soul is immaterial (PL 78/55). Since she thinks that the immaterial can only be something outside of nature (PL 12/14, 187/109), of course an immaterial divine soul would be separable from body; on that point Cavendish has no objection to Descartes. She would, however, insist against Descartes that the divine soul, being supernatural, could not play any role in human life, for that would require that it *interact* with the body—but, she insists, the natural and supernatural cannot interact (PL 11/13). Thus, Cavendish's response to Descartes about the separability of mind and body focuses not on the supernatural or divine soul, but on the "natural" human soul, which for her is the rational animate matter in the human being. Given her claims about the intermixture of all degrees of matter, it follows that in Cavendish's metaphysics, the natural soul is *not* separable from the rest of the human. This, however, is no objection to Descartes, since he never suggests there is anything like Cavendish's "natural" soul.

In letter 36, Cavendish takes on one of Descartes's more notorious claims, that nonhuman animals lack minds. In Part 5 of the *Discourse*, Descartes offers two tests to decide whether a creature has a mind: First, can the creature use language to respond appropriately to all the various questions and comments made to it? Second, can the creature succeed in finding solutions for a broad range of problems? If so, Descartes thinks this is reason to think that the creature is

---

83. Descartes, *Principles of Philosophy*, part 2, section 39.

84. Descartes, *Principles of Philosophy*, part 2, section 33.

85. Daniel Garber, *Descartes' Metaphysical Physics* (Chicago: Chicago University Press, 1992), 220.

employing reason; if not, then Descartes thinks the creature is probably acting only through particular instincts that direct particular types of actions (like nest building, for example).[86]

Descartes had in fact corresponded with Cavendish's husband in 1646 on precisely these questions,[87] so Cavendish perhaps knew of Descartes's views from that source as well. She is unimpressed, characterizing the language argument as "very weak" (PL 113/71). Her first argument against Descartes is not especially compelling: since every part of a human body has knowledge (a view that Descartes would surely deny since he locates knowledge only in the soul), we have reason to think that other creatures have knowledge too (PL 113/71). The reasoning in Cavendish's second argument is stronger, though it reaches a weaker conclusion: that animals do not display the same skills as humans is no evidence that they lack knowledge altogether (PL 114/72). As she reiterates throughout *Philosophical Letters*, not just nonhuman animals but all other parts of nature have knowledge and perception (PL 153/91, 169/100, 184/108, 306–7/148, 517–18/213).

As in her discussion of Hobbes, Cavendish criticizes Descartes's mechanistic account of perception. She considers an example in Descartes's *Optics* of using a stick to perceive objects even in the absence of light or vision. Descartes's point was to provide an analogy for how light can (as he thought) reach a perceiver instantaneously, but Cavendish uses it as a springboard to reiterate her account of "double perception" and patterning (PL 115–16/73–74). The next few letters focus on claims Descartes makes in his *Meteorology* about water, clouds, thunder, and colors. This text is rarely studied today, but it is unsurprising that Cavendish took it seriously, for Descartes's mechanistic, corpuscularian explanations of various natural phenomena were precisely what she sought to oppose with her natural philosophy.[88]

Descartes maintained that matter (itself naturally inert) is divided into small particles (corpuscles) of various shapes and sizes, and that natural phenomena arise from the motions and impacts of these particles on each other, in

---

86. See René Descartes, *Discourse on Method, Optics, Geometry, and Meteorology*, trans. Paul J. Olscamp (Indianapolis: Hackett Publishing Co., 2001).

87. See René Descartes, "To the Marquess of Newcastle, 23 November 1646," in *The Philosophical Writings of Descartes*, vol. 3, trans. John Cottingham, Robert Stoothoff, Dugald Murdoch, and Anthony Kenny (Cambridge: Cambridge University Press, 1991).

88. This essay is included in Olscamp's translation of Descartes's *Discourse*. Craig Martin analyzes the *Meteorology* in *Renaissance Meteorology: Pomponazzi to Descartes* (Baltimore: Johns Hopkins University Press, 2011).

accordance with certain laws of motion. Cavendish objects to both the implausibility of the mechanistic explanations themselves and to the corpuscularianism underlying Descartes's explanations. Against the mechanistic explanations, as we have seen, she thinks they do not adequately explain all the phenomena we observe, and that they sometimes imply that we should see phenomena that do not occur; this is how she argues against Descartes's explanation of thunder, for example (PL 121/75). Against the corpuscularianism, she reiterates that there are no such things as "single" particles (PL 120/74), that is, particles that can really be detached from the rest of matter. Again, her reasoning against the existence of such independent corpuscles relies on inference to the best explanation: if independent corpuscles existed, nature would not be as we observe it to be, namely, orderly and predictable for the most part (PL 431/183); thus corpuscularianism is not the best explanation of natural phenomena.

Henry More

In section 2, Cavendish begins discussing two books by Henry More (1614–1687), *An Antidote against Atheism* (1653) and *The Immortality of the Soul* (1659), both included in *A Collection of Several Philosophical Writings* (1662).[89] While less well known today than Hobbes and Descartes, More wrote more than two dozen books and was extremely influential in his day.[90] A fellow at Christ's College, Cambridge, More's thinking was initially shaped by Platonism.[91] Like Cavendish, his first forays into philosophical writing were in poetry, with his philosophical poem the *Psychodia Platonica* published in 1642. He then began studying Descartes's work, striking up a correspondence with Descartes; while he did not accept Descartes's mechanistic account of matter, he was, like Descartes, a dualist.[92] In 1650 More began corresponding with Anne Conway (1631–1679)

---

89. Studies of Cavendish's engagement with More's ideas include Sarah Hutton, "Margaret Cavendish and Henry More," in *A Princely Brave Woman*, 185–98; Duncan, "Debating Materialism: Cavendish, Hobbes, and More"; and Emma Wilkins, "'Exploding' Immaterial Substances: Margaret Cavendish's Vitalist-Materialist Critique of Spirits," *British Journal for the History of Philosophy* 24, no. 5 (2016): 858–77.

90. Jasper Reid suggests that More was perhaps "*the* most eminent" English philosopher of the seventeenth century. See Jasper Reid, *The Metaphysics of Henry More* (Dordrecht: Springer, 2012), 1.

91. Details on More's life are drawn from Sarah Hutton, "More, Henry," *Oxford Dictionary of National Biography*, online edition (2008).

92. Reid's *The Metaphysics of Henry More* carefully documents shifts in More's views over time, including his views on Cartesian mechanism. The account given here describes views More endorsed in the two books contained in the 1662 volume that Cavendish discusses.

about how to interpret Descartes's *Principles of Philosophy*;[93] Conway would herself go on to write a philosophical treatise, *The Principles of the Most Ancient and Modern Philosophy* (1690).[94]

Cavendish opens by considering More's suggestion that proofs of God's existence are needed to convert atheists; as we saw, Cavendish thinks all parts of nature innately know and love God, meaning there are no genuine atheists (*PL* 137–38/81). But most of her letters about More discuss his objections to the claim that matter can move without the need for anything immaterial. More agreed with Descartes that matter is itself inert, but he disagreed with Descartes regarding the source of the motion that we observe in matter: whereas Descartes (and Hobbes too) thought that motion was the result of impacts and collisions that could be explained in purely materialist terms, More insisted that something immaterial and spiritual is needed to enable matter to move.[95] Whereas Descartes had claimed that immaterial souls differ fundamentally from matter in that only matter is extended, More maintained that immaterial spirits are extended, too. For More, the key difference between spirit and matter hinges, instead, on divisibility and penetrability. While a physical body is divisible into parts,[96] one part cannot *penetrate* another; that is, one physical thing cannot be in the same place at the same time as another physical thing.[97] On the other hand, what is immaterial is indivisible (or, to use his term, "indiscerpible") and able to penetrate, move, and change matter.[98]

Each human soul pervades the human body to which it belongs, and is responsible for mental operations such as understanding as well as for moving the body.[99] Like Plato, More holds that souls preexist the body;[100] the human

---

93. On the More-Conway correspondence about Descartes, see Hutton, *Anne Conway*, 43–44.

94. More mentioned Cavendish's *Poems, and Fancies* and *Philosophical Letters* in a letter to Conway (Hutton, *Anne Conway*, 114). While Cavendish never mentions Conway, she must have known of her, at the very least because More's *Antidote* and *Immortality of the Soul* are dedicated to Conway.

95. See Sarah Hutton, "Margaret Cavendish and Henry More," 188–89.

96. More does insist that there are smallest physical parts in nature that are not further divisible, a view typically associated with atomism and corpuscularianism; unlike atomists such as Gassendi and Charleton, however, he denied the existence of a void, that is, of empty spaces between atoms (*Immortality of the Soul* [1662], book 1, chapter 6, sections 6–7). On More's atomism, see Reid, *The Metaphysics of Henry More*, 44–73.

97. More, *Immortality*, book 1, chapter 3, section 1, and book 1, chapter 5, section 2.

98. More, *Antidote to Atheism* (1662), book 1, chapter 4, section 3, and *Immortality*, book 1, chapter 3, section 1.

99. More, *Antidote*, book 2, chapter 1, section 1.

100. More, *Immortality*, book 2, chapter 12, section 5.

xxxvi   *Introduction*

body is a kind of temporary, terrestrial "vehicle" for the human soul, but before actuating a human body and after the body's death, the human soul can inhabit either air or the "aethereal" realm of heaven.[101] The souls of angels resemble human souls in being able to sense and reason, but they cannot "vitally actuate" a human body; angelic souls can *only* actuate "aerial or aethereal" bodies.[102] Cavendish mocks More on these points, writing that "if the natural human soul should travel through the airy regions, she would at last grow weary, it being so great a journey, except she did meet with the soul of a horse and so ease herself with riding on horseback" (PL 218/123). More also argues that there must be an immaterial "spirit of nature" or "soul of the world" pervading inorganic nature, which is responsible for movements of those bodies; he calls this "the vicarious power of God upon this great automaton, the world."[103]

Cavendish's approach to More has two aspects: she seeks to answer a number of More's specific objections to the view that matter is self-moving, and she raises her own objections to his arguments that immaterial spirits exist. Regarding More's objections to self-moving matter, Cavendish mentions and responds to his complaints that self-moving matter could not produce the order that we see in the world (PL 150–51/89); that if self-moving matter existed, we would expect to see it refusing to be hammered, burned, or otherwise overpowered (PL 154–56/92); that it is unclear what would regulate self-moving matter (PL 160–63/95); that self-moving matter would have to be entirely uniform in its motions (PL 163–64/97); that positing self-moving matter is akin to atheism (PL 164/97); that matter cannot be self-moving because rest is possible (PL 165–66/98); and that matter would not be able to cohere if it were self-moving (PL 166–67/99). She also takes on his related objections to the view that all matter has sense and perception (PL 168–69/100).

Cavendish offers a variety of objections to More's positive thesis that immaterial spirits exist. She objects that if the mind were immaterial (and hence indivisible), then every part of the body to which it is joined should know what is occurring in every other part of that body, which is clearly not the case since "the hand does not know what pain the head feels" (PL 178/105). She thinks More's conception of a substance that is immaterial yet extended is incoherent (PL 186/109), and she objects, using an implicit appeal to Ockham's razor, that

---

101. More, *Immortality*, book 2, chapter 14, section 6.

102. More, *Immortality*, book 1, chapter 8, section 6.

103. More, *Antidote*, book 2, chapter 2, sections 7 and 13; and *Immortality*, book 2, chapter 10, section 2. On More's "spirit of nature" in *Immortality*, see Reid, *The Metaphysics of Henry More*, 329–37.

there is no need to posit immaterial substances since it is as easy for God to make self-moving matter as it would be to make self-moving immaterial spirits (*PL* 194–95/112–13, 199/115). As she asks regarding More's notion of a universal "spirit of nature," "why should it not be as probable that God did give matter a self-moving power to herself as to have made another creature to govern her?" (*PL* 149/88–89). Moreover, echoing a concern Princess Elisabeth had raised in her correspondence with Descartes, immaterial spirits could never interact with anything material, so positing them does not actually help More explain how inert matter moves (*PL* 195/113).

Jan Baptist van Helmont

Section 3 of *Philosophical Letters* is devoted almost entirely to criticisms of someone few have heard of today, Jan Baptist van Helmont (1580–1644).[104] Trained in medicine at the University of Louvain, van Helmont was an alchemist and physician who gained notoriety for his involvement in a debate over the workings of an ointment known as the "weapon-salve" that allegedly cured wounds through application to the weapon that had caused the wound. The debate was about *how* the ointment worked, not whether it worked. Van Helmont's contention that it cured through natural magic fell afoul of Catholic authorities who insisted that the magic of the weapon-salve was satanic; van Helmont was denounced by the Inquisition and put under house arrest for eight years.[105]

Van Helmont's writings were published after his death by his son Franciscus Mercurius van Helmont (1614–c. 1698), who was also a physician and alchemist. The younger van Helmont would go on to meet many important seventeenth-century philosophers[106] and publish his own books, but at the time Cavendish was writing *Philosophical Letters*, he had so far only published his father's writings. These were first published under the title *Ortus medicinae*

---

104. There is little secondary literature on this aspect of Cavendish's work, but see Stephen Clucas, "Margaret Cavendish's Materialist Critique of Van Helmontian Chymistry," *Ambix* 58, no. 1 (2011): 1–12.

105. Walter Pagel, *Joan Baptista van Helmont: Reformer of Science and Medicine* (Cambridge: Cambridge University Press, 1982), 8–13.

106. The younger van Helmont was a friend of Princess Elisabeth of Bohemia and her family (Hutton, *Anne Conway*, 145); of Anne Conway, whom he met in 1670 through Henry More, and whose work he helped publish after her death (Hutton, *Anne Conway*, 161 and 225); and of the Palatine princess Sophie, Leibniz, and John Locke (Stuart Brown, "F. M. van Helmont: His Philosophical Connections and the Reception of His Later Cabbalistic Philosophy," in *Studies in Seventeenth-Century European Philosophy*, ed. M. A. Stewart, 97–116 [Oxford: Clarendon Press, 1997]).

in 1648; this was translated from Latin into English and published in 1662 as *Oriatrike, or Physick Refined.*

The "Physick" in the title refers not to what we call "physics" today but to medicine. Diseases and their cures were taken quite seriously in early modern natural philosophy; even Descartes, whom we read today largely for his metaphysics and epistemology, announced at the end of his *Discourse on Method* that he intended to "devote the rest of my life to nothing other than trying to acquire some knowledge of nature from which we may derive rules in medicine which are more reliable than those we have had up till now."[107] In Cavendish's day, the dominant system for explaining disease and prescribing cures was Galenism. Derived from the work of the Roman physician Claudius Galenus (129–c. 216), Galenism held that the body contains four humors: blood, phlegm, yellow bile, and black bile. Each of these is associated with some combination of the four qualities of hot, cold, moist, and dry. The healthy body has these humors in the right balance; sickness occurs when the balance is upset. Cures aimed to bring the body back into balance, typically on the principle that contraries would cure. For example, if the problem in the body is excessive heat and moistness, the remedy should be to eat or drink something cold and dry.[108] Galenism was the basis of medical training in seventeenth-century England,[109] and even medical practitioners who lacked formal training—apothecaries, barber-surgeons, toothsetters, and housewives who served as healers for their own families—used treatments based on Galenic theory, such as blood-letting and herbal preparations.

However, Galenism had a competitor in the seventeenth century: some practitioners endorsed newer, chemical approaches to medical theory and treatment, such as the use of mercury, arsenic, or antimony, prepared using alchemical processes such as distillation and incineration.[110] Paracelsus (1493–1541) had been among the first to promote these methods (known now as iatrochemistry), and Jan Baptist van Helmont followed. By the mid-seventeenth century, there was widespread interest in iatrochemistry among not just apothecaries but also

---

107. Descartes, *The Philosophical Writings of Descartes*, vol. 1, 51.

108. For a detailed discussion of Galenism, see Owsei Temkin, *Galenism: Rise and Decline of a Medical Philosophy* (Ithaca: Cornell University Press, 1973).

109. Phyllis Allen, "Medical Education in 17th Century England," *Journal of the History of Medicine and Allied Sciences* 1 (1946): 115–43, at 122.

110. For an overview of the debates, see P. M. Rattansi, "The Helmontian-Galenist Controversy in Restoration England," *Ambix* 12, no. 1 (1964): 1–23.

natural philosophers; for example, Antonio Clericuzio has shown how Robert Boyle engaged with the writings of both Paracelsus and van Helmont.[111]

Cavendish does not entirely condemn iatrochemistry, conceding that "in some desperate cases" chemical remedies might be preferable to the traditional herbal ones (PL 285/140), but nonetheless she maintains that "medicines prepared by the art of fire are more poisonous and dangerous than natural drugs" (PL 378/168; see also PL 352/163). Thus Cavendish criticizes van Helmont's love of alchemy, the "art of fire" (PL 281/138). Her objections to alchemy were varied. She thought it was ineffective; some alchemists sought to find a substance—the elusive "philosopher's stone" (PL 284/139)—that would cure all illnesses, or even transform metals into gold, and clearly this had not been found. Given Cavendish's view that portions of matter can transform from one type of entity to another, she had no in-principle objection to the claim that one kind of metal could be transformed into another. However, she did not think that this could be done by the artificial means employed by alchemists. Gold is a "natural creature," but human activities are "arts," which have only "artificial" effects (PL 285/140), so pursuing a process for changing things to gold is pointless. Moreover, alchemy was expensive and time-consuming (PL 284/140). It used up natural materials only in order to promote human ambition; in one of her speeches in *Orations*, Cavendish says alchemists ruin their estates "through covetousness."[112] Moreover, instead of trying to produce something useful, alchemists seek primarily to dominate nature.[113] Finally, Cavendish thought that alchemists deluded themselves into thinking that their experiments revealed the first principles of nature: "for fire and furnaces do often delude the reason, blind the understanding, and make the judgment stagger" (PL 281/138).

Like Paracelsus, van Helmont rejected the Aristotelian view that there were four elements (earth, air, fire, and water). Paracelsus had maintained that salt, sulfur, and mercury were fundamental principles in nature (a view to which Cavendish alludes at PL 286/141), which van Helmont rejected in favor of the view that all things are made ultimately of water.[114] Water can be animated by *semina*, or seeds, which are imbued with a kind of life-force or vital principle

---

111. Antonio Clericuzio, "From van Helmont to Boyle: A Study of the Transmission of Helmontian Chemical and Medical Theories in Seventeenth-Century England," *British Journal for the History of Science* 26 (1993): 303–34.

112. Cavendish, *Orations*, part 1, "An Oration against the Breaking of Peace, with Their Neighbour-Nations."

113. Cavendish, *Worlds Olio*, 340–41.

114. Pagel, *Joan Baptista van Helmont*, 50–51.

that van Helmont called an *archeus*.[115] The *archeus* contains "ideas" or "images" of the thing whose development it is guiding like a blueprint;[116] the development of every animal, plant, and mineral is guided by plans inherent in *archei*. Van Helmont also appealed to the notion of the *archeus* to explain disease: illness occurs not when the humors are out of balance but when a foreign seed infects the body and takes root, causing irritation or even death.[117] Chemical treatments were believed to operate by eliminating the offending seed.[118]

Cavendish criticizes the incomprehensibility of van Helmont's terminology: "he has such strange terms and unusual expressions as may puzzle anybody to apprehend the sense and meaning of them" (PL 234/129). Indeed, some of van Helmont's terminology was entirely novel. One example was *blas*, a kind of motive power to be found in humans as well as the rest of the natural environment, where it was responsible for weather phenomena such as rain, snow, and wind, as well as the motions of the stars.[119] Another term that he coined has persisted; he invented the term *gas*, using it to refer to a highly rarefied water vapor that he believed existed in all things, and which could be released under conditions like heating.[120] Cavendish had no use for these terms, calling them nothing but "chimeras and fancies" (PL 238/131).

However, on the topic of the weapon-salve, Cavendish agreed with van Helmont that its efficacy was due to natural magic, not a supernatural or satanic power: it occurs through the "sympathy" between blood on the weapon and blood in the injured person (PL 290/142). Likewise, she agreed that the magnetic power of a lodestone is due to sympathy between the stone and iron (PL 290/141–42). But Cavendish thought there was nothing mysterious about sympathies between objects; they are simply a variety of natural self-motions whereby one item, perceiving another, has a desire to draw closer to it (PL 297/144).[121] Cavendish is comfortable calling such actions "natural witchcraft,"

---

115. See Allen G. Debus, *The Chemical Philosophy*, vol. 2 (New York: Science History Publications, 1977), 340.

116. Pagel, *Joan Baptista van Helmont*, 24–25 and 97. In some respects van Helmont's *archei* resemble Cavendish's "rational matter," for *archei* are vital forces, capable of imagining and choosing, that are inherent in all material creatures.

117. Debus, *Chemical Philosophy*, 360–61; and Pagel, *Joan Baptista van Helmont*, 142–44.

118. Pagel, *Joan Baptista van Helmont*, 197–98.

119. Pagel, *Joan Baptista van Helmont*, 87–88.

120. Debus, *Chemical Philosophy*, 341.

121. For illuminating background information, see Ann E. Moyer, "Sympathy in the Renaissance," in *Sympathy: A History*, ed. Eric Schliesser (Oxford: Oxford University Press, 2015), 70–101.

maintaining that people take processes to be supernatural or satanic simply because they do not understand how they occur (*PL* 298/145). But she recognizes that it has been women who have borne the brunt of people's fears of what they do not understand (*PL* 298/145).

Indeed, she takes van Helmont to task at several points for his comments about women.[122] When he accuses women of having ideas that lead to madness, she retorts that men have such ideas too, writing slyly that "to mention no other example, some (I will not speak of your author), their writings and strange opinions in philosophy do sufficiently witness it" (*PL* 244/134). To his insistence that men die more often from dangers like shipwrecks and war, Cavendish notes that women, too, are killed in war—by men—and that many die in childbirth (*PL* 313–14/150–51).[123] And she finds his explanation of birthmarks and birth defects absurd. Van Helmont had suggested that a cherry-shaped birthmark on a baby could be explained by a woman's desire to eat cherries while pregnant; her idea of the cherry affects the way the *archeus* of the embryo governs its development. As Cavendish retorts, "I daresay that there have been millions of women which have longed for some or other thing, and have not been satisfied with their desires, and yet their children have never had on their bodies the prints or marks of those things they longed for" (*PL* 277/137).

Cavendish's Other Targets

Cavendish also responds to an assortment of other figures, some of whom are still well known today, such as William Harvey and Galileo. Others, such as Gideon Harvey, were important in the seventeenth century but have been largely forgotten since then. And there are some about whom Cavendish provides too little information to identify.

William Harvey (1578–1657) is best known today for his work explaining the action of the heart and the circulation of blood in *De Motu Cordis* (1628). However, in section 4, letter 2, Cavendish says she has been reading his book on animal generation, or reproduction, *Anatomical Exercitations Concerning the Generation of Living Creatures* (1653). In this book, Harvey describes his observations of hens (of various kinds of birds) and the day-to-day development of birds'

---

122. On this topic, see Broad, "Cavendish, van Helmont, and the Mad Raging Womb."

123. Broad cautions against concluding from Cavendish's criticisms of van Helmont that she was a feminist or proto-feminist, noting that Cavendish endorsed Galenic medicine, which held that women are inferior to men (Broad, "Cavendish, van Helmont, and the Mad Raging Womb," 59–60).

eggs; turning to mammals, he notes that they, too, come from eggs, and he compares the seeds of plants to eggs.[124] From his observations, Harvey deduces various "theorems," generalizations about the process of reproduction. His explanations invoke processes characterized today as "epigenetic," in which the embryo develops successively, in contrast to theories in which the embryo develops all at once or the preformationist theories that would become prominent later in the seventeenth century (which maintain that an embryo is already fully formed in miniature in the egg, and that embryonic and fetal development simply involve a kind of unfolding or enlargement).[125]

Harvey himself emphasized that the novelty of his work was its methodology; he criticized the ancients such as Aristotle and Galen for their failure to make adequate observations or to use "the light of anatomical dissection."[126] But ultimately Harvey's explanation of fetal development was not that different from Aristotle's or Galen's, for all three invoked a kind of formative power.[127] Harvey did not go in for mechanistic explanations, for good reason; if one's only permissible explanatory factors are matter and motion, and the only laws permissible are those that govern matter in motion, it becomes difficult to explain how the matter in the embryo and fetus end up always resembling the structure of the parents, so that, for example, a dog never gives birth to kittens. Traditional nonmechanistic explanations, appealing to forms and final causes, did not face such difficulties.

Cavendish is not troubled by the questions that worried her mechanistic contemporaries about *how* matter can produce an embryo. If matter is self-moving, it is no more difficult to explain fetal development than it is to explain any other natural phenomenon; in this respect, as in her endorsement of the idea that the embryo and fetus develop successively (PL 423–24/180), Cavendish resembles

---

124. William Harvey, *Anatomical Exercitations Concerning the Generation of Living Creatures* (1653), Exercitation 1.

125. See Justin E. H. Smith, *Divine Machines: Leibniz and the Sciences of Life* (Princeton: Princeton University Press, 2011), 170–71.

126. William Harvey, *Anatomical Exercitations*, sig. a5r.

127. Both Aristotle and Galen explained fetal development by invoking a form transmitted through semen. In Aristotle's account, the male transmits the form while the female contributes only matter (*Generation of Animals*, 727b31–33 and 729b19–23). Galen objected that the female must contribute more than matter since offspring can resemble their mothers, but his explanation, too, invoked the transmission of forms from parent to offspring. See Galen, *On Semen*, ed. and trans. Phillip DeLacy (Berlin: Akademie Verlag, 1992), 117–23.

Harvey.[128] However, Cavendish adds a caveat in her adoption of the view that generation involves a transfer of formative power; since she thinks formative power necessarily inheres in matter, this means that matter must also be transferred. This is why her discussion of Harvey opens with a discussion of the transfer of *matter* from parents to offspring.

In section 4, letter 4, Cavendish considers Galileo's *Dialogue Concerning the Two Chief World Systems* (1632),[129] in which three fictional characters debate the evidence for the Copernican view that the earth and other planets orbit the sun, against the Ptolemaic geocentric system. The character of Salviati is Galileo's thinly disguised spokesman; although the dialogue never pronounces Salviati to be correct, Catholic authorities recognized that Galileo had, in the words of the statement he was forced by the Inquisition to sign, "adduced arguments of great cogency" in favor of the Copernican system, and so Galileo was placed under house arrest for the remaining decade of his life.[130]

An English translation of Galileo's *Dialogue* was included in John Salusbury's *Mathematical Collections and Translations in Two Tomes* (1661),[131] but the quotations in *Philosophical Letters* do not match those in Salusbury's edition. Presumably Cavendish had someone translate from the Italian for her, just as she had someone translate Descartes from the Latin. She criticizes various claims about motion put forth by Salviati in the first dialogue; although Salviati was arguing against the geocentric model of the solar system that drew on Aristotelian principles, he agreed with Aristotle regarding the perfection of circular motion, a claim Cavendish rejects. Cavendish does not, however, engage with the arguments for which the *Dialogue* is best known, such as its evidence that the other planets orbit the sun and that the moon orbits the earth, its discussions of problems with the Ptolemaic system, and debates that Cavendish would presumably have found interesting about whether the earth is a giant lodestone, whether there are sympathies and antipathies in nature, and what causes the tides.

Letter 8 takes up the views of "that learned author Dr. Ch." (*PL* 451/194). This is Walter Charleton (1620–1707), author of *Physiologia*

---

128. See Benjamin Goldberg, "Epigenesis and the Rationality of Nature in William Harvey and Margaret Cavendish," *History and Philosophy of the Life Sciences* 39 (2017), article no. 8.

129. See Galileo Galilei, *Dialogue Concerning the Two Chief World Systems*, trans. Stillman Drake (New York: Modern Library, 2001).

130. Stillman Drake, "The Translator's Preface," in *Dialogue Concerning the Two Chief World Systems*, xxxiv.

131. John Salusbury, *Mathematical Collections and Translations in Two Tomes* (1661).

Epicuro-Gassendo-Charltoniana: Or, A Fabrick of Science Natural, Upon the Hypothesis of Atoms (1654).[132] As the title indicates, Charleton endorsed an atomistic natural philosophy, derived from the views of French philosopher Pierre Gassendi (1592–1655). Charleton argued that matter is not infinitely divisible; rather, nature is composed of naturally indivisible particles, or atoms.[133] There are also void spaces, or vacua, into and out of which atoms can move. Atomism was first expressed by the ancient philosophers Leucippus, Democritus, and Epicurus, and was also a theme of Lucretius's poem *De Rerum Natura*, written sometime in the first century BCE; the rediscovery of Lucretius's work in 1417 led to a revival of atomism in western Europe, particularly in the mid-seventeenth century.[134]

Cavendish rejects the existence of both atoms and vacuum. She reads Charleton as maintaining that a vacuum is an incorporeal extension, which she says is contradictory: being noncorporeal, it must be nothing, yet being extended, it must have dimensions; but how can a nothing have dimensions (PL 451/194)? Regarding atoms, Cavendish reiterates her views that "there is no part in nature individable, no not that so small a part which the Epicureans name an 'atom'" and that "there can never be a single part by itself" (PL 455/196). Oddly, she quotes some lines from *Poems, and Fancies* (1653) to show "the ground" of her opinions about atoms (PL 455/196–97). *Poems, and Fancies* includes numerous poems about how the motions of atoms of various shapes produce such entities as earth, air, fire, and water. Her quotations from these poems in *Philosophical Letters* might suggest she still endorses these views; however, it is unclear that Cavendish ever actually *endorsed* atomism in the first place. Atomism may have been for her merely a fertile topic on which to exercise her creative imagination, just as she wrote poems on fairies—especially since she described the poems as "fictions," writing that "fiction is not given for truth, but pastime."[135] And even if Cavendish did once believe atomism was true, she repudiated it in *Philosophical*

---

132. Charleton and Cavendish actually corresponded; Charleton's three extant letters can be found in Broad, *Women Philosophers of Seventeenth-Century England*. The letters from Cavendish are no longer extant.

133. Walter Charleton, *Physiologia Epicuro-Gassendo-Charltoniana: Or, A Fabrick of Science Natural, Upon the Hypothesis of Atoms* (1654), book 2, chapter 2.

134. On the development of seventeenth-century atomism, see Antonio Clericuzio, "The Atomism of the Cavendish Circle," *The Seventeenth Century* 9, no. 2 (1994): 247–73; and Reid, "Atoms and Void," in *The Metaphysics of Henry More*, 35–43. On Cavendish's views on atomism, see Karen Detlefsen, "Atomism, Monism, and Causation in the Natural Philosophy of Margaret Cavendish," in *Oxford Studies in Early Modern Philosophy*, vol. 3, ed. Daniel Garber and Steven Nadler, 199–240 (Oxford: Clarendon Press, 2006); and Boyle, *The Well-Ordered Universe*, 40–61.

135. Cavendish, "To Natural Philosophers," in *Poems, and Fancies*, sig. a6r.

and Physical Opinions (1655), in the preface entitled "A Condemning Treatise of Atoms." Moreover, as we have seen, she offers philosophical arguments against the possibility of both vacuum and atoms. The inclusion of passages from her atomistic poems should thus not be read as an indication that she meant those passages literally.

Cavendish also discusses the works of several writers left unnamed. In section 4, letter 10, she refers to "the book of your new author that treats of natural philosophy, which I perceive is but lately come forth" (PL 456/197), without naming either the author or the book. However, the quotations come from Gideon Harvey's *Archelogia Philosophica Nova, or, New Principles of Philosophy* (1663). This Harvey (1636/7–1702)—no relation to William Harvey—was a Dutch physician who settled in London in the 1650s and went on to write numerous books on diseases.[136] *Archelogia* was his first publication; ranging from metaphysics and epistemology to cosmogony and astronomy, Harvey's book sought to provide explanations of natural beings and phenomena in terms of the four elements of earth, air, fire, and water, which themselves were to be explained in terms of *minima*, essentially atoms. As in her discussion of van Helmont, Cavendish criticizes Harvey for his "hard, intricate, and nonsensical words" (PL 456/197). Indeed, Harvey seems to have invented such terms as *catochization* (a word that does not even appear in the comprehensive *Oxford English Dictionary*) and *ontology*[137] (now widely used).

Cavendish raises brief objections to a passage in Harvey's text where he maintains (against van Helmont) that inanimate matter cannot be a "co-agent" with active matter,[138] and to his claims that there are *minima* and *maxima* in nature. But she focuses most on passages addressing the issue with which she began *Philosophical Letters*: whether a material universe can be infinite in extension and in time. Harvey maintains that because the universe is finite in its parts, the universe as a whole must also be finite; similarly, since particular creatures have beginnings and endings, nature itself must have a beginning and ending.[139] Cavendish notes that these arguments are invalid; one cannot infer anything about the whole simply from features of its parts (PL 457/198).

---

136. Patrick Wallis, "Harvey, Gideon," in *Oxford Dictionary of National Biography*, online edition (2008).

137. "ontology, n.," *Oxford English Dictionary Online*, (June 2004; revised September 2019).

138. Gideon Harvey, *Archelogia Philosophica Nova, or, New Principles of Philosophy* (1663), part 2, book 1, chapter 5, section 3.

139. See Harvey, *Archelogia*, part 2, book 1, chapter 9, section 8.

Cavendish is much more positive about an unnamed source that she considers in letter 22, where she addresses "the works of that learned and ingenious writer B" (*PL* 495/202). Given her subsequent reference to "his experiments," Cavendish is probably referring to Robert Boyle (1627–1691), although her comments are too vague to indicate which of his works she has in mind. The well-educated son of the wealthy first Earl of Cork, Boyle had become interested in natural philosophy in 1649, when he set up a laboratory in his home for performing alchemical experiments and making microscopic observations.[140] In Oxford during the 1650s, Boyle continued his experimental work, and in 1660 he helped found the Royal Society of London, the institution dedicated to scientific study that Cavendish would visit in 1667. There she saw Boyle demonstrate the use of the famous "air pump" that he, Robert Hooke, and others used to study the possibility of a vacuum in nature; she also observed some chemical experiments showing how colors and temperature could be changed in mixtures and a demonstration involving a sixty-pound lodestone.[141]

In that same letter, Cavendish mentions another book, *A General Collection of Discourses of the Virtuosi of France* (1664).[142] This book was a compilation of one hundred of the weekly "conferences" held between 1633 and 1642 at the Bureau d'Adresse in Paris, an institute founded in 1630 by Théophraste Renaudot. The conferences were open to the general public and included discussions and debates on natural philosophy, inventions, navigation, medicine and diet, law, literature, moral philosophy, and education.[143] Cavendish zeroes in on two of the more practical topics covered in the book, writing that she wished she could learn two "arts" mentioned there: "to argue without error in all kinds, modes, and figures in a quarter of an hour," and "to learn a way to understand all languages in six hours" (*PL* 497/203).[144]

---

140. Michael Hunter, "Boyle, Robert" in *Oxford Dictionary of National Biography*, online edition (2008).

141. See Samuel Mintz, "The Duchess of Newcastle's Visit to the Royal Society," *Journal of English and Germanic Philology* 51 (1952): 168–76. Mintz's claims about Cavendish are inaccurate—for example, he unaccountably insists that Cavendish "had no regard for Bacon's dictum about knowledge 'for benefit and use'" (168)—but he provides interesting details about the visit itself.

142. Anonymous, *A General Collection of Discourses of the Virtuosi of France, upon Questions of All Sorts of Philosophy and Other Natural Knowledge [...] Rendered into English by G. Havers* (1664).

143. See Howard Solomon, "The Conferences at the Bureau d'Adresse," in *Public Welfare, Science and Propaganda in 17th-Century France: The Innovations of Théophraste Renaudot* (Princeton: Princeton University Press, 1972), 60–64.

144. See *General Collection of Discourses of the Virtuosi of France*, part 2, conferences 25 and 26.

Some of the sources that Cavendish mentions are difficult to identify now. For example, she brings up the debate over free will and predestination in section 4, letter 26, where she says she was recently discussing these topics with "Sir. P. H. and Sir R. L." (*PL* 503/205). Who could these be? Since she refers to *conversing* with these men rather than to books they published, they may not have been authors at all, making it difficult even to guess at their identity. Indeed, Sir P. H. and Sir R. L. may not be real people; in Cavendish's *CCXI Sociable Letters*, also published in 1664, the fictional persona of the letter writer uses initials to refer to many people who also seem to be merely fictional.[145] Nonetheless, the debate between Bishop Bramhall and Thomas Hobbes on free will and necessity, conducted in 1645 at William Cavendish's home and then carried on in print into the next decade, attests to the contemporary importance of these issues. Cavendish's reference to it is further evidence of her attunement to her philosophical peers, an attunement that is clear throughout *Philosophical Letters* even as she uses those letters to explain and promote her own philosophical system.

---

145. On the identification of characters mentioned in *CCXI Sociable Letters*, see Lara Dodds, "Reading and Writing in *Sociable Letters*: Or, How Margaret Cavendish Read Her Plutarch," *English Literary Renaissance* 41, no. 1 (2011): 189–218, especially 197.

A NOTE ON THE TEXT

This is an abridged version of *Philosophical Letters*, containing 107 of the original 157 letters. In general, I have abridged or omitted letters that either (1) simply repeat but do not further elaborate on material that Cavendish has addressed in another letter; or (2) focus mainly on replying to someone else's views without saying much about Cavendish's *own* philosophical views. I also have omitted some of the book's front matter: a dedicatory poem by William Cavendish, a preface justifying the book in the form of a letter from Margaret to William, and a preface addressed to the University of Cambridge.

Section 1, covering Thomas Hobbes and René Descartes, includes thirty-five of Cavendish's forty-five letters. Section 2, on Henry More, includes twenty-seven out of thirty-four letters. Section 3, on Jan Baptist van Helmont, contains twenty-five (many abridged) out of forty-five letters. Section 4, which contains letters covering a variety of lesser-known figures and some letters clarifying some points from the 1663 *Philosophical and Physical Opinions*, includes twenty out of thirty-three letters.

To make the letters more accessible to twenty-first-century readers, I have modernized Cavendish's writing in several ways. First, capitalization, use of apostrophes, and spelling have been modernized; for example, I have changed archaic spellings such as "shew" to "show," "spake" to "spoke," and "doth" to "does." However, I have retained archaic terms or terms that Cavendish coined (such as "composable" and "deitical"), explaining their meanings in footnotes when necessary. All definitions are drawn from the Oxford English Dictionary Online (Oxford University Press, June 2020). In cases where I needed to supply a missing word or phrase to make sense of Cavendish's meaning, my additions are enclosed in square brackets.

Second, I have removed Cavendish's italics, except for book titles, replacing these with quotation marks. I have also added quotation marks to show when Cavendish is mentioning (rather than using) a term. Third, I have freely changed Cavendish's punctuation. I have not attempted to bring her punctuation completely in line with modern usage, because doing so would have required substantial changes to her sentence structures. However, to improve readability, I have removed many of Cavendish's commas, added some, and sometimes substituted commas for semicolons or colons. When Cavendish used a semicolon followed by a capital letter, or used a colon where we would use a period, I have substituted a period. Finally, I have added paragraph breaks, since Cavendish rarely used them.

When Cavendish used marginal notes to provide information about the texts she quoted, I have provided this information in footnotes. In cases when she did not provide information, I have sought to identify the source. When she refers to views expressed in her own *Philosophical and Physical Opinions*, I have tried to identify the specific passages to which she was likely referring.

The source used for this edition was a microfilm reproduction of *Philosophical Letters* made from an original in Cambridge University Library. The original pagination is given in square brackets. Omitted passages are indicated by ellipses.

MARGARET CAVENDISH

Her Life, Her Times

1623	Margaret Lucas is born near Colchester, Essex, the youngest of eight children of Thomas Lucas and Elizabeth Leighton.
1625	Death of Thomas Lucas.
	Charles I becomes king and begins a costly war with Spain.
1629–1640	King Charles I dissolves Parliament, makes peace with Spain, and rules for eleven years without Parliamentary approval, raising money through a series of unpopular taxes.
1637	King Charles I attempts to force the Church of Scotland to adopt practices of the Church of England.
	René Descartes publishes *Discourse on Method*, with the essays *Optics*, *Meteorology*, and *Geometry*, in the Netherlands.
1639	The Scottish Army invades England.
1640	Short on cash to fight the Scots, Charles calls Parliament back, dissolves it again when members of Parliament seek to limit his powers, and reconvenes it at the end of the year.
1640–1642	Tensions rise between Charles and Parliament, and between Royalists and Parliamentarian supporters in the countryside.
1641	Descartes publishes *Meditations on First Philosophy*.
1642	Local anti-Royalist residents outside Colchester plunder the Lucas home.
	The first battle of the English Civil War is fought at Edgehill. King Charles moves the court to Oxford.
1643	Margaret joins the court of Queen Henrietta Maria, the French wife of Charles I, in Oxford.

1644	Scottish troops join the English Civil War on the side of the Parliamentarians.
	William Cavendish, Marquess of Newcastle, leads the king's troops in a losing battle at Marston Moor. After the defeat, Cavendish moves to the Netherlands.
	Queen Henrietta Maria and her entourage escape England for France.
	Descartes publishes *Principles of Philosophy*, intended as a textbook.
1645	William Cavendish moves to Paris to join the queen's court.
	William Cavendish hosts a debate between Hobbes and Bramhall on the topic of free will.
	William and Margaret marry.
1648	The Cavendishes move to Rotterdam, Netherlands. Shortly thereafter, they move to Antwerp, renting the former home of Peter Paul Rubens.
1649	King Charles I is executed.
	Parliament establishes the Commonwealth of England to replace the monarchy.
1651	Margaret goes to England with her brother-in-law Charles Cavendish to petition Parliament (ultimately unsuccessfully) for access to William's money.
	Thomas Hobbes publishes *Leviathan*.
1653	Oliver Cromwell becomes "Lord Protector" of the Commonwealth.
	Margaret publishes *Poems, and Fancies* and *Philosophicall Fancies* in London.
	Henry More publishes *An Antidote against Atheism*, with a second edition in 1655; it is later included in *A Collection of Several Philosophical Writings* (1662).
1654	Walter Charleton publishes his exposition of atomism, *Physiologia Epicuro-Gassendo-Charltoniana*.

1655	Margaret publishes *Worlds Olio* and *Philosophical and Physical Opinions* in London.
1656	Margaret publishes *Natures Pictures*, containing stories and her autobiography.
	An English translation of Hobbes's *De Corpore* (1655), entitled *Elements of Philosophy, the First Section, Concerning Body*, is published.
1658	Death of Oliver Cromwell.
1659	Henry More publishes *The Immortality of the Soul*, later included in *A Collection of Several Philosophical Writings* (1662).
1660	Restoration of King Charles II.
	The Royal Society of London is started, with a royal charter granted in 1663.
	The Cavendishes return to England, settling at Welbeck.
1662	Margaret publishes *Plays* and *Orations of Divers Sorts*.
	An English translation of Jan Baptist van Helmont's *Oriatrike, or Physick Refined* is published.
1663	Margaret publishes the second, revised, edition of *Philosophical and Physical Opinions*.
1664	Margaret publishes *CCXI Sociable Letters* and *Philosophical Letters*, plus the second edition of *Poems, and Fancies*.
1665	William and Margaret are titled the Duke and Duchess of Newcastle.
1666	Margaret publishes *Observations upon Experimental Philosophy* and *Description of a New World, Called the Blazing World*.
	The plague ravages London.
1667	Margaret tours the Royal Society of London.
	Margaret publishes *The Life of the Thrice Noble, High and Puissant Prince William Cavendishe, Duke, Marquess and Earl of Newcastle* in an attempt to restore his reputation.
1668	Margaret publishes *Plays, Never before Printed*; *Grounds of Natural Philosophy*; and second editions of *Observations* and *Blazing World*.

1671   Margaret publishes second editions of *Natures Pictures* and *Worlds Olio*.

1673   Death of Margaret (December 15).

1674   Burial of Margaret at Westminster Abbey.

1676   William publishes *Letters and Poems in Honour of the Incomparable Princess, Margaret, Duchess of Newcastle*.

Death of William, buried at Westminster Abbey with Margaret.

SUGGESTIONS FOR FURTHER READING

Secondary Literature

Battigelli, Anna. *Margaret Cavendish and the Exiles of the Mind*. Lexington: University Press of Kentucky, 1998.

Boyle, Deborah. *The Well-Ordered Universe: The Philosophy of Margaret Cavendish*. New York: Oxford University Press, 2018.

Broad, Jacqueline. "Margaret Cavendish." In *Women Philosophers of the Seventeenth Century*, 35–64. Cambridge: Cambridge University Press, 2002.

Clericuzio, Antonio. "Chemistry and Atomism in England (1600–1660)." In *Elements, Principles and Corpuscles: A Study of Atomism and Chemistry in the Seventeenth Century*, 75–102. Dordrecht: Kluwer, 2000.

Clucas, Stephen, ed. *A Princely Brave Woman: Essays on Margaret Cavendish, Duchess of Newcastle*. Aldershot: Ashgate, 2003.

Cottegnies, Line, and Nancy Weitz, eds. *Authorial Conquests: Essays on Genre in the Writings of Margaret Cavendish*. Madison, NJ: Fairleigh Dickinson University Press, 2003.

Cunning, David. *Cavendish*. London: Routledge, 2016.

Cunning, David. "Margaret Lucas Cavendish." In *Stanford Encyclopedia of Philosophy* online, edited by Edward N. Zalta. Article published October 16, 2009; last modified May 29, 2017; https://plato.stanford.edu/archives/sum2017/entries/margaret-cavendish.

Detlefsen, Karen. "Reason and Freedom: Margaret Cavendish on the Order and Disorder of Nature." *Archiv für Geschichte der Philosophie* 89, no. 2 (2007): 157–91.

Garber, Daniel. *Descartes' Metaphysical Physics*. Chicago: Chicago University Press, 1992.

Hattab, Helen. "The Mechanical Philosophy." In *The Oxford Handbook of Philosophy in Early Modern Europe*, edited by Desmond M. Clarke and Catherine Wilson, 71–95. Oxford: Oxford University Press, 2011.

James, Susan. "The Philosophical Innovations of Margaret Cavendish." *British Journal for the History of Philosophy* 7, no. 2 (1999): 219–44.

Jones, Kathleen. *A Glorious Fame: The Life of Margaret Cavendish, Duchess of Newcastle, 1623–1673*. London: Bloomsbury, 1988.

Kargon, Robert Hugh. *Atomism in England from Hariot to Newton*. Oxford: Clarendon Press, 1966.

LoLordo, Antonia. *Pierre Gassendi and the Birth of Early Modern Philosophy*. New York: Cambridge University Press, 2006.

Mendelson, Sara H., ed. *Ashgate Critical Essays on Women Writers in England, 1550–1700*. Vol. 7. Farnham, Surrey: Ashgate, 2009.

Pagel, Walter. *Joan Baptista van Helmont: Reformer of Science and Medicine*. Cambridge: Cambridge University Press, 1982.

Reid, Jasper. *The Metaphysics of Henry More*. Dordrecht: Springer, 2012.

Sarasohn, Lisa T. *The Natural Philosophy of Margaret Cavendish: Reason and Fancy during the Scientific Revolution*. Baltimore: Johns Hopkins University Press, 2010.

Siegfried, Brandie R., and Lisa T. Sarasohn, eds. *God and Nature in the Thought of Margaret Cavendish*. Farnham, Surrey: Ashgate, 2014.

Walters, Lisa. *Margaret Cavendish: Gender, Science and Politics*. Cambridge: Cambridge University Press, 2014.

Whitaker, Katie. *Mad Madge: The Extraordinary Life of Margaret Cavendish, Duchess of Newcastle, the First Woman to Live by Her Pen*. New York: Basic Books, 2002.

Primary Literature: Margaret Cavendish

Bowerbank, Sylvia, and Sara Mendelson, eds. *Paper Bodies: A Margaret Cavendish Reader*. Peterborough, Ontario: Broadview Press, 2000.

Cavendish, Margaret. *Grounds of Natural Philosophy*. Edited by Anne Thell. Peterborough, Ontario: Broadview Press, 2020.

Cavendish, Margaret. *Observations upon Experimental Philosophy*. Edited by Eileen O'Neill. Cambridge: Cambridge University Press, 2001.

Cavendish, Margaret. *Observations upon Experimental Philosophy, Abridged*. Edited by Eugene Marshall. Indianapolis: Hackett Publishing Co., 2016.

Cavendish, Margaret. *Philosophical and Physical Opinions (1663 edition): A Digital Edition*. Edited by Marcy P. Lascano. https://cavendish-ppo.ku.edu/.

Cavendish, Margaret. *Poems and Fancies with The Animal Parliament*. Edited by Brandie R. Siegfried. Tempe, AZ: Iter Press, 2018.

Cavendish, Margaret. *Sociable Letters*. Edited by James Fitzmaurice. Peterborough, Ontario: Broadview Press, 2004.

Cunning, David, ed. *Margaret Cavendish: Essential Writings*. New York: Oxford University Press, 2019.

James, Susan, ed. *Margaret Cavendish: Political Writings*. Cambridge: Cambridge University Press, 2003.

Primary Literature: Cavendish's Targets in *Philosophical Letters*

Charleton, Walter. *Physiologia Epicuro-Gassendo-Charltoniana*. Index and introduction by Robert Hugh Kargon. New York: Johnson Reprint Co., 1966.

Descartes, René. *Discourse on Method, Optics, Geometry, and Meteorology*. Translated by Paul J. Olscamp. Indianapolis: Hackett Publishing Co., 2001.

Descartes, René. *The Philosophical Writings of Descartes*. Translated by John Cottingham, Robert Stoothoff, and Dugald Murdoch. Vol. 1. Cambridge: Cambridge University Press, 1985.

Descartes, René. *The Philosophical Writings of Descartes*. Translated by John Cottingham, Robert Stoothoff, Dugald Murdoch, and Anthony Kenny. Vol. 3. Cambridge: Cambridge University Press, 1991.

Galilei, Galileo. *Dialogue Concerning the Two Chief World Systems*. Translated and with revised notes by Stillman Drake. New York: Modern Library, 2001.

Harvey, William. *Anatomical Exercitations, Concerning the Generation of Living Creatures.* New York: Classics of Medicine Library, 1991.

Hobbes, Thomas. *Elements of Philosophy: The First Section, Concerning Body.* In *The English Works of Thomas Hobbes,* edited by Sir William Molesworth. Vol. 1. London: John Bohn, 1839.

Hobbes, Thomas. *Leviathan.* Edited by Edwin Curley. Indianapolis: Hackett Publishing Co., 1994.

More, Henry. *A Collection of Several Philosophical Writings, 1662.* 2 vol. New York: Garland Publishing, 1978.

Stewart, M. A., ed. *Selected Philosophical Papers of Robert Boyle.* Indianapolis: Hackett Publishing Co., 1991.

Philosophical Letters

A PREFACE TO THE READER

[b1r] Worthy Readers,

I did not write this book out of delight, love, or humor of contradiction; for I would rather praise than contradict any person or persons that are ingenious; but by reason opinion is free and may pass without a passport, I took the liberty to declare my own opinions as other philosophers do, and to that purpose I have here set down several famous and learned authors' opinions, and my answers to them in the form of letters, which was the easiest way for me to write; and by so doing, I have done that which I would have done unto me; for I am as willing to have my opinions contradicted as I do contradict others. For I love reason so well that whosoever can bring most rational and probable arguments shall have my vote, although [b1v] against my own opinion.

But you may say, if contradictions were frequent, there would be no agreement amongst mankind. I answer, it is very true; wherefore contradictions are better in general books than in particular families, and in schools better than in public states, and better in philosophy than in divinity. All which considered, I shun, as much as I can, not to discourse or write of either church or state. But I desire so much favor, or rather justice of you, worthy readers, as not to interpret my objections or answers any other ways than against several opinions in philosophy; for I am confident there is not any body that does esteem, respect, and honor learned and ingenious persons more than I do. Wherefore judge me neither to be of a contradicting humor nor of a vain-glorious mind for dissenting from other men's opinions, but rather that it is done out of love to truth, and to make my own opinions the more intelligible, which cannot better be done than by arguing and comparing other men's opinions with them.

The authors whose opinions I mention, I have read as I found them printed in my native language, except Descartes, who being in Latin, I had some few places translated to me out of his works; and I must confess that since I have read the works of these learned men, I understand the names and terms of art a little better than I did before; but it is not so much as to make me a scholar, nor yet so little but that had I read more before I did begin to write my other book called *Philosophical Opinions*, they would have been more intelligible; for

3

my error was, I began to write so early that I had not lived [b2r] so long as to be able to read many authors.[1] I cannot say, I divulged my opinions as soon as I had conceived them, but yet I divulged them too soon to have them artificial and methodical. But since what is past cannot be recalled, I must desire you to excuse those faults, which were committed for want of experience and learning. As for school-learning,[2] had I applied myself to it, yet I am confident I should never have arrived to any; for I am so incapable of learning that I could never attain to the knowledge of any other language but my native, especially by the rules of art. Wherefore I do not repent that I spent not my time in learning, for I consider, it is better to write wittily than learnedly; nevertheless, I love and esteem learning, although I am not capable of it.

But you may say, I have expressed neither wit nor learning in my writings. Truly, if not, I am the more sorry for it; but self-conceit, which is natural to mankind, especially to our sex, did flatter and secretly persuade me that my writings had sense and reason, wit and variety; but judgment being called not to counsel, I yielded to self-conceit's flattery, and so put out my writings to be printed as fast as I could, without being reviewed or corrected. Neither did I fear any censure, for self-conceit had persuaded me I should be highly applauded; wherefore I made such haste that I had three or four books printed presently after each other.

But to return to this present work, I must desire you, worthy readers, to read first my book called *Philosophical and Physical Opinions*, before you censure this, for this book is but an explanation of the former, wherein is contained the ground of my opinions, and those [b2v] that will judge well of a building must first consider the foundation; to which purpose I will repeat some few heads and principles of my opinions, which are these following:

First, that nature is infinite and the eternal servant of God. Next, that she is corporeal and partly self-moving, dividable, and composable; that all and every particular creature, as also all perception and variety in nature, is made by corporeal self-motion, which I name sensitive and rational matter, which is life and knowledge, sense and reason. Again, that these sensitive and rational

---

1. The first edition of *Philosophical and Physical Opinions*, an expansion of *Philosophicall Fancies* (1653), was published in 1655. The second edition, greatly revised, was published in 1663. Cavendish typically refers to this book simply as *Philosophical Opinions*.

2. This is a reference to Aristotelian scholastic philosophy, so called because its practitioners were often monks who taught in universities and colleges. By the time Cavendish was writing her works, the highly technical philosophical terminology, distinctions, and debates of the "schools" were beginning to come under fire by critics such as Descartes and Robert Boyle.

parts of matter are the purest and subtlest parts of nature, as the active parts, the knowing, understanding, and prudent parts, the designing, architectonical,[3] and working parts, nay, the life and soul of nature, and that there is not any creature or part of nature without this life and soul; and that not only animals, but also vegetables, minerals and elements, and what more is in nature are endued with this life and soul, sense and reason. And because this life and soul is a corporeal substance, it is both dividable and composable; for it divides and removes parts from parts, as also composes and joins parts to parts, and works in a perpetual motion without rest; by which actions not any creature can challenge[4] a particular life and soul to itself, but every creature may have by the dividing and composing nature of this self-moving matter more or fewer natural souls and lives.

These and the like actions of corporeal nature or natural matter you may find more at large described in my aforementioned book of *Philosophical Opinions*, and more clearly repeated and explained in this present. [c1r] 'Tis true, the way of arguing I use is common, but the principles, heads, and grounds of my opinions are my own, not borrowed or stolen in the least from any; and the first time I divulged them was in the year 1653, since which time I have reviewed, reformed, and reprinted them twice; for at first, as my conceptions were new and my own, so my judgment was young and my experience little, so that I had not so much knowledge as to declare them artificially and methodically; for as I mentioned before, I was always unapt to learn by the rules of art.

But although they may be defective for want of terms of art and artificial expressions, yet I am sure they are not defective for want of sense and reason. And if anyone can bring more sense and reason to disprove these my opinions, I shall not repine or grieve, but either acknowledge my error, if I find myself in any, or defend them as rationally as I can, if it be but done justly and honestly, without deceit, spite, or malice; for I cannot choose but acquaint you, noble readers, I have been informed, that if I should be answered in my writings, it would be done rather under the name and cover of a woman than of a man; the reason is because no man dare or will set his name to the contradiction of a lady; and to confirm you the better herein, there has one chapter of my book called *The Worlds Olio*, treating of a monastical life, been answered already in a little pamphlet under the name of a woman, although she did little towards it; wherefore it being a hermaphroditical book, I judged it not worthy of taking notice

---

3. Architectonical: directive, controlling.
4. Challenge: lay claim to.

of.[5] The like I shall do to any other that will answer this present work of mine or contradict my opinions indirectly with fraud and deceit.

[c1v] But I cannot conceive why it should be a disgrace to any man to maintain his own or others' opinions against a woman, so it be done with respect and civility; but to become a cheat by dissembling, and quit the breeches for a petticoat, merely out of spite and malice, is base and not fit for the honor of a man or the masculine sex. Besides, it will easily be known; for a philosopher or philosopheress is not produced on a sudden. Wherefore, although I do not care, nor fear contradiction, yet I desire it may be done without fraud or deceit, spite and malice; and then I shall be ready to defend my opinions the best I can, whilst I live, and after I am dead I hope those that are just and honorable will also defend me from all sophistry, malice, spite, and envy, for which Heaven will bless them.

In the meantime, worthy readers, I should rejoice to see that my works are acceptable to you, for if you be not partial, you will easily pardon those faults you find when you do consider both my sex and breeding; for which favor and justice, I shall always remain,

> Your most obliged servant,
> M. N.

---

5. This text is Suzanne du Verger's *Humble Reflections* (1657). See Kathryn Coad Narramore, "Du Verger's *Humble Reflections* and *Dedicatory Epistles* as Public Sphere," *Prose Studies: History, Theory, Criticism* 35 (2013): 139–53; and Stewart Duncan, "The Letters in the *Philosophical Letters*" (updated April 2014), https://stewartduncan.org/letters-philosophical-letters/.

SECTION 1

1.

Madam,

You have been pleased to send me the works of four famous learned authors, to wit, of two most famous philosophers of our age, Descartes and Hobbes, and of that learned philosopher and divine Dr. More, as also of that famous physician and chemist van Helmont. Which works you have sent me not only to peruse, but also to give my judgment of them, and to send you word by the usual way of our correspondence, which is by letters, how far, and wherein I do differ from these famous authors, their opinions in natural philosophy.[1]

To tell you truly, Madam, your commands did at first much affright me, for it did appear as if you had commanded me to get upon a high rock and fling myself into the sea, [2] where neither a ship nor a plank nor any kind of help was near to rescue me and save my life; but that I was forced to sink, by reason I cannot swim.

So I, having no learning nor art to assist me in this dangerous undertaking, thought I must of necessity perish under the rough censures of my readers, and be not only accounted a fool for my labor, but a vain and presumptuous person, to undertake things surpassing the ability of my performance; but on the other side I considered first, that those worthy authors, were they my censurers, would not deny me the same liberty they take themselves; which is that I may dissent from their opinions as well as they dissent from others, and from amongst themselves. And if I should express more vanity than wit, more ignorance than knowledge, more folly than discretion, it being according to the nature of our sex, I hoped that my masculine readers would civilly excuse me, and my female readers could not justly condemn me.

Next I considered with myself that it would be a great advantage for my book called *Philosophical Opinions*, as to make it more perspicuous and intelligible by

---

1. Natural philosophy: natural science.

the opposition of other opinions, since two opposite things placed near each other are the better discerned; for I must confess that when I did put forth my philosophical work at first, I was not so well skilled in the terms or expressions usual in natural philosophy; and therefore for want of their knowledge, I could not declare my meaning so plainly and clearly as I ought to have done, which may be a sufficient argument to my readers that I have not read heretofore any natural philosophers, and taken some light from them; but that my opinions did merely [3] issue from the fountain of my own brain, without any help or assistance.

Wherefore, since for want of proper expressions, my named book of philosophy was accused of obscurity and intricacy, I thought your commands would be a means to explain and clear it the better, although not by an artificial way as by logical arguments or mathematical demonstrations, yet by expressing my sense and meaning more properly and clearly than I have done heretofore. But the chief reason of all was the authority of your command, which did work so powerfully with me that I could not resist, although it were to the disgrace of my own judgment and wit; and therefore I am fully resolved now to go on as far and as well as the natural strength of my reason will reach.

But since neither the strength of my body, nor of my understanding or wit, is able to mark every line or every word of their works, and to argue upon them, I shall only pick out the ground opinions of the aforementioned authors, and those which do directly dissent from mine, upon which I intend to make some few reflections according to the ability of my reason; and I shall merely go upon the bare ground of natural philosophy and not mix divinity with it, as many philosophers use to do, except it be in those places, where I am forced by the author's arguments to reflect upon it, which yet shall be rather with an expression of my ignorance than a positive declaration of my opinion or judgment thereof; for I think it not only an absurdity but an injury to the holy profession of divinity to draw her to the proofs in natural philosophy; wherefore I shall strictly follow the guidance of natural reason and keep to my own ground and principles as much as I can; [4] which that I may perform the better, I humbly desire the help and assistance of your favor, that according to that real and entire affection you bear to me, you would be pleased to tell me unfeignedly if I should chance to err or contradict but the least probability of truth in any thing; for I honor truth so much, as I bow down to its shadow with the greatest respect and reverence; and I esteem those persons most that love and honor truth with the same zeal and fervor, whether they be ancient or modern writers.

Thus, Madam, although I am destitute of the help of arts, yet being supported by your favor and wise directions, I shall not fear any smiles of scorn or words of reproach; for I am confident you will defend me against all the mischievous and poisonous teeth of malicious detractors. I shall, besides, implore the assistance of the sacred church and the learned schools to take me into their protection and shelter my weak endeavors. For though I am but an ignorant and simple woman, yet I am their devoted and honest servant, who shall never quit the respect and honor due to them, but live and die theirs, as also, Madam,

Your Ladyship's humble and faithful servant.[2]

[5] **2.**

Madam,

Before I begin my reflections upon the opinions of those authors you sent me, I will answer first your objection concerning the ground of my philosophy, which is infinite matter. For you were pleased to mention that you could not well apprehend how it was possible that many infinites could be contained in one infinite, since one infinite takes up all place imaginary,[3] leaving no room for any other. Also, if one infinite should be contained in another infinite, that which contains must of necessity be bigger than that which is contained, whereby the nature of infinite would be lost, as having no bigger nor less, but being of an infinite quantity.

First of all, Madam, there is no such thing as all in infinite, nor any such thing as all the place, for infinite is not circumscribed nor limited. Next, as for that one infinite cannot be in another infinite, I answer, as well as one finite can be in another finite; for one creature is not only composed of parts, but one part lies within another, and one figure within another, and one motion within another. As for example, animal kind, have they not internal and external parts, and so internal and external motions? And are not animals, vegetables, and minerals enclosed in the elements?

---

2. Signing a letter with the phrase "your humble servant" or "your obedient servant" was commonplace in seventeenth-century correspondence.

3. Imaginary: imaginable.

But as for infinites, you must know, Madam, that there are several kinds of infinites. For there is, first, infinite in quantity [6] or bulk, that is such a big and great corporeal substance, which exceeds all bounds and limits of measure, and may be called "infinite in magnitude." Next there is infinite in number, which exceeds all numeration and account, and may be termed "infinite in multitude." Again there is infinite in quality; as for example, infinite degrees of softness, hardness, thickness, thinness, heat and cold, etc.; also infinite degrees of motion, and so infinite creations, infinite compositions, dissolutions, contractions, dilations, digestions, expulsions; also infinite degrees of strength, knowledge, power, etc. Besides there is infinite in time, which is properly named "eternal."

Now, when I say that there is but one infinite, and that infinite is the only matter, I mean "infinite in bulk and quantity." And this only matter, because it is infinite in bulk, must of necessity be divisible into infinite parts, that is, infinite in number, not in bulk or quantity; for though infinite parts in number make up one infinite in quantity, yet they considered in themselves cannot be said "infinite," because every part is of a certain limited and circumscribed figure, quantity, and proportion, whereas infinite has no limits nor bounds. Besides, it is against the nature of a single part to be infinite, or else there would be no difference between the part and the whole, the nature of a part requiring that it must be less than its whole, but all what is less has a determined quantity, and so becomes finite. Therefore it is no absurdity to say that an infinite may have both finite and infinite parts, finite in quantity, infinite in number.

But those that say, if there were an infinite body, that each of its parts must of necessity be infinite too, are much mistaken; for it is a contradiction [7] in the same terms to say "one infinite part," for the very name of a part includes a finiteness; but take all parts of an infinite body together, then you may rightly say they are infinite. Nay, reason will inform you plainly; for example, imagine an infinite number of grains of corn in one heap; surely if the number of grains be infinite, you must grant of necessity the bulk or body, which contains this infinite number of grains, to be infinite too; to wit, infinite in quantity, and yet you will find each grain in itself to be finite.

But you will say, an infinite body cannot have parts, for if it be infinite, it must be infinite in quantity, and therefore of one bulk and one continued[4] quantity, but infinite parts in number make a discrete quantity. I answer, it is all one; for a body of a continued quantity may be divided and severed into so many parts either actually, or mentally in our conceptions or thoughts; besides, nature

---

4. Continued: continuous, without breaks or gaps.

is one continued body, for there is no such vacuum in nature, as if her parts did hang together like a linked chain; nor can any of her parts subsist single and by itself, but all the parts of infinite nature, although they are in one continued piece, yet are they several[5] and discerned[6] from each other by their several figures. And by this, I hope you will understand my meaning when I say that several infinites may be included or comprehended in one infinite; for by the "one infinite," I understand infinite in quantity, which includes infinite in number, that is, infinite parts; then infinite in quality, as infinite degrees of rarity, density, swiftness, slowness, hardness, softness, etc., infinite degrees of motions, infinite creations, dissolutions, contractions, dilations, alterations, etc., [8] infinite degrees of wisdom, strength, power, etc.; and lastly infinite in time or duration, which is eternity, for infinite and eternal are inseparable. All which infinites are contained in the only matter as many letters are contained in one word, many words in one line, many lines in one book.

But you will say, perhaps, if I attribute an infinite wisdom, strength, power, knowledge, etc. to nature, then nature is in all coequal with God, for God has the same attributes. I answer, not at all; for I desire you to understand me rightly when I speak of infinite nature, and when I speak of the infinite deity, for there is great difference between them, for it is one thing a deitical[7] or divine infinite, and another a natural infinite. You know that God is a spirit and not a bodily substance; again, that nature is a body and not a spirit; and therefore none of these infinites can obstruct or hinder each other, as being different in their kinds, for a spirit being no body, requires no place, place being an attribute which only belongs to a body, and therefore when I call nature "infinite," I mean an infinite extension of body, containing an infinite number of parts; but what[8] does an infinite extension of body hinder the infiniteness of God, as an immaterial spiritual being?

Next, when I do attribute an infinite power, wisdom, knowledge, etc. to nature, I do not understand a divine, but a natural infinite wisdom and power, that is, such as properly belongs to nature, and not a supernatural, as is in God. For nature, having infinite parts of infinite degrees, must also have an infinite natural wisdom to order her natural infinite parts and actions, and consequently an infinite natural power to put her wisdom [9] into act; and so the rest of her

---

5. Several: separate.

6. Discerned: differentiated.

7. By "deitical" (a term she seems to have coined herself), Cavendish means "pertaining to God."

8. What: in what way.

attributes, which are all natural. But God's attributes, being supernatural, transcend much these natural infinite attributes; for God, being the God of nature, has not only nature's infinite wisdom and power, but besides, a supernatural and incomprehensible infinite wisdom and power; which in no ways do hinder each other, but may very well subsist together. Neither do God's infinite justice and his infinite mercy hinder each other; for God's attributes, though they be all several infinites, yet they make but one infinite.

But you will say, if nature's wisdom and power extends no further than to natural things, it is not infinite, but limited and restrained. I answer, that does not take away the infiniteness of nature; for there may be several kinds of infinites, as I related before, and one may be as perfect an infinite as the other in its kind. For example: suppose a line to be extended infinitely in length, you will call this line "infinite," although it have not an infinite breadth. Also, if an infinite length and breadth join together, you will call it an "infinite superficies,"[9] although it wants[10] an infinite depth; and yet every infinite, in its kind, is a perfect infinite, if I may call it so.

Why then shall not nature also be said to have an infinite natural wisdom and power, although she has not a divine wisdom and power? Can we say, man has not a free will, because he has not an absolute free will, as God has? Wherefore, a natural infinite and the infinite God may well stand together, without any opposition or hindrance, or without any detracting or derogating from the omnipotency and glory of God; for God remains still the God of nature, [10] and is an infinite immaterial purity, when as[11] nature is an infinite corporeal substance; and immaterial and material cannot obstruct each other. And though an infinite corporeal cannot make an infinite immaterial, yet an infinite immaterial can make an infinite corporeal, by reason there is as much difference in the power as in the purity. And the disparity between the natural and the divine infinite is such as they cannot join, mix, and work together, unless you do believe that divine actions can have allay.[12]

But you may say, purity belongs only to natural things, and none but natural bodies can be said purified, but God exceeds all purity. 'Tis true. But if there were infinite degrees of purity in matter, matter might at last become immaterial, and so from an infinite material turn to an infinite immaterial, and from

---

9. Superficies: surface.

10. Wants: lacks.

11. When as: whereas.

12. Allay: alloy, mixture.

nature to be God: a great, but an impossible change. For I do verily believe that there can be but one omnipotent God, and he cannot admit of addition or diminution; and that which is material cannot be immaterial, and what is immaterial cannot become material, I mean, so as to change their natures; for nature is what God was pleased she should be; and will be what she was, until God be pleased to make her otherwise.

Wherefore there can be no new creation of matter, motion, or figure; nor any annihilation of any matter, motion, or figure in nature, unless God do create a new nature. For the changing of matter into several particular figures does not prove an annihilation of particular figures, nor the cessation of particular motions an annihilation of them. Neither does the variation of the only matter produce an annihilation of any [11] part of matter, nor the variation of figures and motions of matter cause an alteration in the nature of only matter. Wherefore there cannot be new lives, souls, or bodies in nature; for, could there be any thing new in nature, or any thing annihilated, there would not be any stability in nature, as a continuance of every kind and sort of creatures, but there would be a confusion between the new and old matter, motions, and figures, as between old and new nature. In truth, it would be like new wine in old vessels, by which all would break into disorder. Neither can supernatural and natural effects be mixed together, no more than material and immaterial things or beings. Therefore it is probable, God has ordained nature to work in herself by his leave, will, and free gift.

But there have been, and are still, strange and erroneous opinions and great differences amongst natural philosophers concerning the principles of natural things; some will have them "atoms";[13] others will have the first principles to be salt, sulfur, and mercury;[14] some will have them to be the four elements, as fire, air, water, and earth;[15] and others will have but one of these elements; also some will have "gas" and "blas," "ferments," "ideas," and the like;[16] but what they believe to be principles and causes of natural things are only effects; for in all probability it appears to human sense and reason that the cause of every particular material creature is the only and infinite matter, which has motions and figures inseparably united; for matter, motion, and figure are but one thing, individable in its nature.

And as for immaterial spirits, there is surely no such thing in infinite nature, to wit, so as to be parts of nature; for nature [12] is altogether material, but this

---

13. On seventeenth-century atomists and Cavendish's relation to atomism, see the Introduction (p. lxiv–xlv).

14. I.e., Swiss physician and alchemist Paracelsus (1493–1541).

15. I.e., Aristotelian natural philosophers.

16. I.e., alchemist Jan Baptist van Helmont (1580–1644), whose views Cavendish criticizes in section 3. See the Introduction (p. xxxvii–xli).

opinion proceeds from the separation or abstraction of motion from matter, *viz.*, that man thinks matter and motion to be dividable from each other, and believes motion to be a thing by itself, naming it an immaterial thing, which has a being, but not a bodily substance. But various and different effects do not prove a different matter or cause, neither do they prove an unsettled cause, only the variety of effects has obscured the cause from the several parts, which makes particular creatures partly ignorant and partly knowing. But in my opinion, nature is material, and not any thing in nature, what belongs to her, is immaterial; but whatsoever is immaterial is supernatural. Therefore motions, forms, thoughts, ideas, conceptions, sympathies, antipathies, accidents, qualities, as also natural life and soul, are all material. And as for colors, scents, light, sound, heat, cold, and the like, those that believe them not to be substances or material things, surely their brain or heart (take what place you will for the forming of conceptions) moves very irregularly, and they might as well say our sensitive organs are not material; for what objects soever that are subject to our senses cannot in sense be denied to be corporeal, when as those things that are not subject to our senses can be conceived in reason to be immaterial.

But some philosophers, striving to express their wit, obstruct reason; and drawing [on] divinity to prove sense and reason, weaken faith so, as their mixed divine philosophy becomes mere poetical fictions and romancical expressions, making material bodies immaterial spirits, and immaterial spirits material bodies; and some have conceived some [13] things to be neither material nor immaterial, but between both. Truly, Madam, I wish their wits had been less and their judgments more, as not to jumble natural and supernatural things together, but to distinguish either clearly, for such mixtures are neither natural nor divine. But as I said, the confusion comes from their too nice abstractions, and from the separation of figure and motion from matter, as not conceiving them individable; but if God and his servant nature were as intricate[17] and confuse[18] in their works as men in their understandings and words, the universe and production of all creatures would soon be without order and government, so as there would be a horrid and eternal war both in heaven and in the world; and so pitying their troubled brains and wishing them the light of reason that they may clearly perceive the truth, I rest, Madam,

Your real friend and faithful servant.

---

17. Intricate: entangled.
18. Confuse: confusedly mixed.

3.

Madam,

It seems you are offended at my opinion that nature is eternal without beginning, which, you say, is to make her God, or at least coequal with God. But if you apprehend my meaning rightly, you will [14] say, I do not. For first, God is an immaterial and spiritual infinite being, which propriety[19] God cannot give away to any creature, nor make another God in essence like to him, for God's attributes are not communicable to any creature. Yet this does not hinder that God should not make infinite and eternal matter, for that is as easy to him as to make a finite creature, infinite matter being quite of another nature than God is, to wit, corporeal, when God is incorporeal, the difference whereof I have declared in my former letter.

But as for nature, that it cannot be eternal without beginning, because God is the creator and cause of it, and that the creator must be before the creature, as the cause before the effect, so that it is impossible for nature to be without a beginning; if you will speak naturally, as human reason guides you, and bring an argument concluding from the priority of the cause before the effect, give me leave to tell you that God is not tied to natural rules, but that he can do beyond our understanding, and therefore he is neither bound up to time, as to be before; for if we will do this, we must not allow that the eternal son of God is coeternal with the father, because nature requires a father to exist before the son, but in God is no time, but all eternity; and if you allow that God has made some creatures, as supernatural spirits, to live eternally, why should he not as well have made a creature from all eternity? For God's making is not our making; he needs no priority of time.

But you may say, the comparison of the eternal generation of the Son of God is mystical and divine, and not to be applied to natural things. I answer, the action by which God created the world[20] or made [15] nature, was it natural or supernatural? Surely you will say it was a supernatural and God-like action; why then will you apply natural rules to a God-like and supernatural action? For what man knows how and when God created nature?

You will say, the Scripture does teach us that, for it is not six thousand years, when God created this world. I answer, the holy Scripture informs us only of

---

19. Propriety: property, quality.
20. Cavendish often uses "world" to mean the whole created cosmos, not just Earth.

the creation of this visible world, but not of nature and natural matter; for I firmly believe according to the word of God, that this world has been created as is described by Moses, but what is that to natural matter? There may have been worlds before, as many are of the opinion that there have been men before Adam, and many amongst divines[21] do believe that after the destruction of this world God will create a new world again, as a new heaven and a new earth; and if this be probable, or at least may be believed without any prejudice to the holy Scripture, why may it not be probably believed that there have been other worlds before this visible world? For nothing is impossible with God; and all this does derogate nothing from the honor and glory of God, but rather increases his divine power. But as for the creation of this present world, it is related that there was first a rude[22] and indigested[23] heap, or chaos, without form, void and dark; and God said, "Let it be light. Let there be a firmament in the midst of the waters, and let the waters under the heaven be gathered together, and let the dry land appear. Let the Earth bring forth grass, the herb yielding seed, and the fruit-tree yielding fruit after its own kind; and let there be lights in the firmament, the one to rule the day and [16] the other the night; and let the waters bring forth abundantly the moving creature that has life; and let the Earth bring forth living creatures after its kind; and at last God said, Let us make man, and all what was made, God saw it was good."[24] Thus all was made by God's command, and who executed his command but the material servant of God, nature? Which ordered her self-moving matter into such several figures as God commanded, and God approved of them.

And thus, Madam, I verily believe the creation of the world, and that God is the sole and omnipotent creator of heaven and earth and of all creatures therein; nay, although I believe nature to have been from eternity, yet I believe also that God is the God and author of nature, and has made nature and natural mat-ter in a way and manner proper to his omnipotency and incomprehensible by us. I will pass by natural arguments and proofs, as not belonging to such an omnipotent action; as for example, how the nature of relative terms requires that they must both exist at one point of time, *viz.*, a master and his servant, and a king and his subjects; for one bearing relation to the other can in no ways be considered as different from one another in formiliness[25] or laterness of time;

---

21. Divines: theologians.

22. Rude: rough, unfinished, unformed.

23. Indigested: without order or arrangement.

24. Gen. 1:3–31 (abridged).

25. Formiliness: earliness.

but as I said, these being merely natural things, I will nor cannot apply them to supernatural and divine actions.

But if you ask me, how is it possible that nature, the effect and creature of God, can be eternal without beginning? I will desire you to answer me first, how a creature can be eternal without end, as, for example, supernatural spirits are, and then I will answer you, how a creature can be eternal without beginning. [17] For eternity consists herein, that it has neither beginning nor end; and if it be easy for God to make a being without end, it is not difficult for him to make a being without beginning. One thing more I will add, which is that if nature has not been made by God from all eternity, then the title of God as being a creator, which is a title and action upon which our faith is grounded (for it is the first article in our creed) has been accessory[26] to God, as I said, not full six thousand years ago; but there is not any thing accessory to God, he being perfection itself.

But, Madam, all what I speak is under the liberty of natural philosophy, and by the light of reason only, not of revelation; and my reason being not infallible, I will not declare my opinions for an infallible truth. Neither do I think that they are offensive either to church or state; for I submit to the laws of one, and believe the doctrine of the other, so much, that if it were for the advantage of either, I should be willing to sacrifice my life, especially for the church; yea, had I millions of lives, and every life was either to suffer torment or to live in ease, I would prefer torment for the benefit of the church; and therefore, if I knew that my opinions should give any offense to the church, I should be ready every minute to alter them. And as much as I am bound in all duty to the obedience of the church, as much am I particularly bound to your Ladyship, for your entire love and sincere affection towards me, for which I shall live and die, Madam,

Your most faithful friend, and humble servant.

[18] **4.**

Madam,

I have chosen, in the first place, the work of that famous philosopher Hobbes, called *Leviathan*, wherein I find he says, "That the cause of sense or sensitive perception is the external body or object, which presses the organ proper to

---

26. Accessory: incidental.

each sense."[27] To which I answer, according to the ground of my own *Philosophical Opinions*, that all things, and therefore outward objects as well as sensitive organs, have both sense and reason,[28] yet neither the objects nor the organs are the cause of them; for perception is but the effect of the sensitive and rational motions, and not the motions of the perception; neither does the pressure of parts upon parts make perception; for although matter by the power of self-motion is as much composable as dividable, and parts do join to parts, yet that does not make perception; nay, the several parts, between which the perception is made, may be at such a distance as not capable to press.

As for example, two men may see or hear each other at a distance, and yet there may be other bodies between them that do not move to those perceptions, so that no pressure can be made, for all pressures are by some constraint and force; wherefore, according to my opinion, the sensitive and rational free motions do pattern out each other's object, as figure and voice in each other's eye and ear; for life and knowledge, which I name rational and sensitive matter, are in every creature and [19] in all parts of every creature, and make all perceptions in nature, because they are the self-moving parts of nature, and according as those corporeal, rational, and sensitive motions move, such or such perceptions are made. But these self-moving parts being of different degrees (for the rational matter is purer than the sensitive), it causes a double perception in all creatures, whereof one is made by the rational corporeal motions, and the other by the sensitive; and though both perceptions are in all the body and in every part of the body of a creature, yet the sensitive corporeal motions having their proper organs, as work-houses, in which they work some sorts of perceptions, those perceptions are most commonly made in those organs, and are double again; for the sensitive motions work either on the inside or on the outside of those organs, on the inside in dreams, on the outside awake; and although both the rational and the sensitive matter are inseparably joined and mixed together, yet do they not always work together, for oftentimes the rational works without any sensitive patterns, and the sensitive again without any rational patterns.

But mistake me not, Madam, for I do not absolutely confine the sensitive perception to the organs, nor the rational to the brain, but as they are both in the whole body, so they may work in the whole body according to their own motions. Neither do I say that there is no other perception in the eye but sight,

---

27. Thomas Hobbes, *Leviathan* (1651), part 1, chapter 1. The phrase "or sensitive perception" is Cavendish's addition.

28. See *Philosophical and Physical Opinions* (1663), part 1, chapter 17; and part 4, chapter 8.

in the ear but hearing, and so forth, but the sensitive organs have other perceptions besides these; and if the sensitive and rational motions be irregular in those parts between which the perception is made, as for example, in the two aforementioned men that see and hear each other, [20] then they both neither see nor hear each other perfectly; and if one's motions be perfect, but the other's irregular and erroneous, then one sees and hears better than the other; or if the sensitive and rational motions move more regularly and make perfecter patterns in the eye than in the ear, then they see better than they hear; and if more regularly and perfectly in the ear than in the eye, they hear better than they see. And so it may be said of each man singly, for one man may see the other better and more perfectly than the other may see him; and this man may hear the other better and more perfectly than the other may hear him; whereas if perception were made by pressure, there would not be any such mistakes; besides, the hard pressure of objects, in my opinion, would rather annoy and obscure, than inform.

But as soon as the object is removed, the perception of it made by the sensitive motions in the organs ceases, by reason the sensitive motions cease from patterning, but yet the rational motions do not always cease so suddenly, because the sensitive corporeal motions work with the inanimate matter, and therefore cannot retain particular figures long, whereas the rational matter does only move in its own substance and parts of matter, and upon none other, as my book of *Philosophical Opinions* will inform you better.[29] And thus perception, in my opinion, is not made by pressure, nor by species,[30] nor by matter going either from the organ to the object, or from the object into the organ. By this it is also manifest that understanding comes not from exterior objects or from the exterior sensitive organs; for as exterior objects do not make perception, so they do neither make [21] understanding, but it is the rational matter that does it, for understanding may be without exterior objects and sensitive organs. And this in short is the opinion of, Madam,

Your faithful friend and servant.

---

29. Cavendish discusses the motions of rational matter in *Philosophical and Physical Opinions* (1663), part 1, chapter 24; and part 2, chapter 12.

30. Cavendish here alludes to the Aristotelian theory of perception embraced by many scholastic philosophers: an object is perceived when the "sensible species" of that object emanates from it and enters the sense organ and mind of the perceiver.

5.

Madam,

Our author's opinion is that "when a thing lies still, unless somewhat else stir it, it will lie still forever; but when a thing is in motion, it will eternally be in motion, unless somewhat else stay it; the reason is," says he, "because nothing can change itself."[31] To tell you truly, Madam, I am not of his opinion, for if matter moves itself, as certainly it does, then the least part of matter, were it so small as to seem individable, will move itself. 'Tis true, it could not desist from motion, as being its nature to move, and nothing can change its nature; for God himself, who has more power than self-moving matter, cannot change himself from being God; but that motion should proceed from another exterior body joining with or touching that body which it moves, is in my opinion not probable; for though nature is all corporeal, and her actions are corporeal motions, yet that does not prove that the motion of particular [22] creatures or parts is caused by the joining, touching, or pressing of parts upon parts; for it is not the several parts that make motion, but motion makes them; and yet motion is not the cause of matter, but matter is the cause of motion, for matter might subsist without motion, but not motion without matter, only there could be no perception without motion, nor no variety, if matter were not self-moving; but matter, if it were all inanimate and void of motion, would lie as a dull, dead, and senseless heap.

But that all motion comes by joining or pressing of other parts, I deny, for if sensitive and rational perceptions, which are sensitive and rational motions in the body and in the mind, were made by the pressure of outward objects pressing the sensitive organs, and so the brain or interior parts of the body, they would cause such dents and holes therein as to make them sore and patched in a short time. Besides, what was represented in this manner would always remain, or at least would not so soon be dissolved, and then those pressures would make a strange and horrid confusion of figures, for not any figure would be distinct. Wherefore my opinion is that the sensitive and rational matter do make or pattern out the figures of several objects, and do dissolve them in a moment of time; as for example, when the eye sees the object first of a man, then of a horse, then of another creature, the sensitive motions in the eye move first into the figure of the man, then straight into the figure of the horse, so that the man's figure is dissolved and altered into the figure of the horse, and so forth; but if the eye

---

31. Hobbes, *Leviathan*, part 1, chapter 2.

sees many figures at once, then so many several figures are made by the sensitive corporeal motions, [23] and as many by the rational motions, which are sight and memory, at once.

But in sleep both the sensitive and rational motions make the figure without patterns, that is, exterior objects, which is the cause that they are often erroneous, whereas if it were the former impression of the objects, there could not possibly be imperfect dreams or remembrances, for fading of figures requires as much motion as impression, and impression and fading are very different and opposite motions;[32] nay, if perception was made by impression, there could not possibly be a fading or decay of the figures printed either in the mind or body, whereas yet, as there is alteration of motions in self-moving matter, so there is also an alteration of figures made by these motions.

But you will say, it does not follow, if perception be made by impression, that it must needs continue and not decay; for if you touch and move a string, the motion does not continue for ever, but ceases by degrees. I answer, there is a great difference between prime self-motion, and forced or artificial motions; for artificial motions are only an imitation of natural motions, and not the same, but caused by natural motions; for although there is no art that is not made by nature, yet nature is not made by art. Wherefore we cannot rationally judge of perception by comparing it to the motion of a string, and its alteration to the ceasing of that motion, for nature moves not by force but freely. 'Tis true, 'tis the freedom in nature for one man to give another a box on the ear, or to trip up his heels, or for one or more men to fight with each other; yet these actions are not like the actions of loving embraces and kissing each other; neither are the [24] actions one and the same when a man strikes himself and when he strikes another; and so is likewise the action of impression and the action of self-figuring not one and the same, but different; for the action of impression is forced, and the action of self-figuring is free.

Wherefore the comparison of the forced motions of a string, rope, watch, or the like can have no place here; for though the rope, made of flax or hemp, may have the perception of a vegetable, yet not of the hand, or the like, that touched or struck it; and although the hand does occasion the rope to move in such a manner, yet it is not the motion of the hand by which it moves, and when it ceases, its natural and inherent power to move is not lessened; like as a man that has left off carving or painting has no less skill than he had before, neither is that

---

32. Hobbes maintains that the mind retains (in memory) faint images of objects it has perceived, but that these images fade over time. See *Leviathan*, part 1, chapter 2.

skill lost when he plays upon the lute or virginals,[33] or plows, plants, and the like, but he has only altered his action, as from carving to painting, or from painting to playing, and so to plowing and planting, which is not through disability but choice.

But you will say, it is nevertheless a cessation of such a motion. I grant it. But the ceasing of such a motion is not the ceasing of self-moving matter from all motions, neither is cessation as much as annihilation, for the motion lies in the power of the matter to repeat it as oft it will, if it be not overpowered, for more parts, or more strength, or more motions may overpower the less. Wherefore forced or artificial and free natural motions are different in their effects, although they have but one cause, which is the self-moving matter, and though matter is but active and passive, yet there is great variety, and so great difference in force and liberty, objects and [25] perceptions, sense and reason, and the like.

But to conclude, perception is not made by the pressure of objects, no more than hemp is made by the rope-maker, or metal by the bell-sounder or ringer, and yet neither the rope nor the metal is without sense and reason, but the natural motions of the metal and the artificial motions of the ringer are different; wherefore a natural effect in truth cannot be produced from an artificial cause, neither can the ceasing of particular forced or artificial motions be a proof for the ceasing of general, natural, free motions, as that matter itself should cease to move; for there is no such thing as rest in nature, but there is an alteration of motions and figures in self-moving matter, which alteration causes variety as well in opinions, as in everything else.

Wherefore in my opinion, though sense alters, yet it does not decay, for the rational and sensitive part of matter is as lasting as matter itself, but that which is named decay of sense is only the alteration of motions, like as the motions of memory and forgetfulness, and the repetition of the same motions is called "remembrance." And thus much of this subject for the present, to which I add no more but rest, Madam,

Your faithful friend and servant.

---

33. Virginals: a small harpsichord.

[26]   **6.**

Madam,

Your author, discoursing of imagination, says, "That as soon as any object is removed from our eyes, though the impression that is made in us remains, yet other objects more present succeeding and working on us, the imagination of the past is obscured and made weak."[34] To which I answer, first, that he conceives sense and imagination to be all one, for he says, "imagination is nothing else, but a fading or decaying sense";[35] whereas in my opinion they are different, not only their matter but their motions also being distinct and different; for imagination is a rational perception, and sense a sensitive perception; wherefore as much as the rational matter differs from the sensitive, as much does imagination differ from sense.

Next I say that impressions do not remain in the body of sensitive matter, but it is in its power to make or repeat the like figures. Neither is imagination less when the object is absent than when present, but the figure patterned out in the sensitive organs, being altered and remaining only in the rational part of matter, is not so perspicuous and clear as when it was both in the sense and in the mind. And to prove that imagination of things past does not grow weaker by distance of time, as your author says, many a man in his old age will have as perfect an imagination of what is past in his younger years as if he saw it present.

And as for your author's opinion that "imagination and memory are one and the same,"[36] I grant that they are made [27] of one kind of matter; but although the matter is one and the same, yet several motions in the several parts make imagination and memory several things. As for example, a man may imagine that which never came into his senses, wherefore imagination is not one and the same thing with memory. But your author seems to make all sense, as it were, one motion, but not all motion sense, whereas surely there is no motion but is either sensitive or rational; for reason is but a pure and refined sense, and sense a grosser reason.

Yet all sensitive and rational motions are not one and the same; for forced or artificial motions, though they proceed from sensitive matter, yet are they so different from the free and prime natural motions that they seem, as it were,

---

34. Hobbes, *Leviathan*, part 1, chapter 2.

35. Hobbes, *Leviathan*, part 1, chapter 2. In this and most of her quotations from other authors, Cavendish has slightly altered the text. I have only noted her alterations when they change the original author's meaning.

36. Hobbes, *Leviathan*, part 1, chapter 2.

quite of another nature. And this distinction neglected is the cause that many make appetites and passions, perceptions and objects, and the like, as one, without any or but little difference. But having discoursed of the difference of these motions in my former letter, I will not be tedious to you with repeating it again, but remain, Madam,

Your faithful friend and servant.

[28]   **7.**

Madam,

Your author's opinion concerning dreams seems to me in some part very rational and probable, in some part not. For when he says that "dreams are only imaginations of them that sleep, which imaginations have been before either totally or by parcels in the sense; and that the organs of sense, as the brain and the nerves, being benumbed in sleep, as not easily to be moved by external objects, those imaginations proceed only from the agitation of the inward parts of man's body, which for the connection they have with the brain and other organs, when they be distempered, so keep the same in motion, whereby the imaginations there formerly made, appear as if a man were waking,"[37] this seems to my reason not very probable.

For, first, dreams are not absolutely imaginations, except we do call all motions and actions of the sensitive and rational matter "imaginations." Neither is it necessary that all imaginations must have been before either totally or by parcels in the sense; neither is there any benumbing of the organs of sense in sleep. But dreams, according to my opinion, are made by the sensitive and rational corporeal motions, by figuring several objects, as awake; only the difference is that the sensitive motions in dreams work by rote[38] and on the inside of the sensitive organs, when as awake they work according to the patterns of outward objects; and exteriously or on the outside of the [29] sensitive organs, so that sleep or dreams are nothing else but an alteration of motions, from

---

37. Hobbes, *Leviathan*, part 1, chapter 2.
38. In Cavendish's day, as in ours, to do something "by rote" typically meant to do it mechanically and without understanding. However, in Cavendish's idiosyncratic usage, it describes the motions of sensitive and rational matter when they act creatively, without copying something else.

moving exteriously to mov[ing] interiously, and from working after a pattern to work[ing] by rote.

I do not say that the body is without all exterior motions when asleep, as breathing and beating of the pulse (although these motions are rather interior than exterior), but that only the sensitive organs are outwardly shut, so as not to receive the patterns of outward objects; nevertheless the sensitive motions do not cease from moving inwardly, or on the inside of the sensitive organs. But the rational matter does often, as awake, so asleep or in dreams, make such figures as the sensitive did never make either from outward objects or of its own accord; for the sensitive has sometimes liberty to work without objects, but the rational much more, which is not bound either to the patterns of exterior objects, or of the sensitive voluntary figures.

Wherefore it is not diverse distempers, as your author says, that cause different dreams, or cold, or heat; neither are dreams the reverse of our waking imaginations, nor all the figures in dreams are not made "with their heels up, and their heads downward,"[39] though some are; but this error or irregularity proceeds from want of exterior objects or patterns, and by reason the sensitive motions work by rote; neither are the motions reverse, because they work inwardly asleep and outwardly awake, for mad-men awake see several figures without objects.

In short, sleeping and waking is somewhat after that manner, when men are called either out of their doors, or stay within their houses; or like a ship where the mariners work [30] all under hatches, whereof you will find more in my *Philosophical Opinions*;[40] and so taking my leave, I rest, Madam,

Your faithful friend and servant.

8.

Madam,

Your author, going on in his discourse of imagination, says "that, as we have no imagination, whereof we have not formerly had sense, in whole or in parts;

---

39. That is, upside-down. Hobbes says that dreams are "the reverse of our waking imaginations," meaning that the images in dreams result from motions of the "inward parts of man's body," while images formed by the imagination when awake derive from motions of objects outside the body (*Leviathan*, part 1, chapter 2).

40. Cavendish discusses sleep and dreams in *Philosophical and Physical Opinions* (1663), part 6, chapters 19–24; for the ship analogy, see chapter 22.

so we have not transition from one imagination to another, whereof we never had the like before in our senses."[41] To which my answer is, in short, that the rational part of matter is one composed figure, as in man or the like creature may make such figures as the senses did never make in that composed figure or creature. And though your author reproves those that say "imaginations rise of themselves";[42] yet if the self-moving part of matter which I call rational makes imaginations, they must needs rise of themselves; for the rational part of matter, being free and self-moving, depends upon nothing, neither sense nor object, I mean, so as not to be able to work without them.

Next, when your author, defining "understanding," says that it is nothing else but "an imagination raised by words or [31] other voluntary signs,"[43] my answer is that understanding, and so words and signs, are made by self-moving matter, that is, sense and reason, and not sense and reason by words and signs; wherefore thoughts are not like "water upon a plain table, which is drawn and guided by the finger this or that way,"[44] for every part of self-moving matter is not always forced, persuaded, or directed, for if all the parts of sense and reason were ruled by force or persuasion, not any wounded creature would fail to be healed, or any disease to be cured by outward applications, for outward applications to wounds and diseases might have more force than any object to the eye. But though there is great affinity and sympathy between parts, yet there is also great difference and antipathy between them, which is the cause that many objects cannot with all their endeavors work such effects upon the interior parts, although they are closely pressed, for impressions of objects do not always affect those parts they press.

Wherefore, I am not of your author's opinion that all parts of matter press one another. It is true, Madam, there cannot be any part single, but yet this does not prove that parts must needs press each other. And as for his "train of thoughts," I must confess that thoughts for the most part are made orderly, but yet they do not follow each other like geese, for surely man has sometimes very different thoughts; as for example, a man sometime is very sad for the death of his friend, and thinks of his own death, and immediately thinks of a wanton mistress, which later thought, surely, the thought of death did not draw in;

---

41. Hobbes, *Leviathan*, part 1, chapter 3.
42. Hobbes, *Leviathan*, part 1, chapter 2.
43. Hobbes, *Leviathan*, part 1, chapter 2.
44. Hobbes, *Leviathan*, part 1, chapter 3.

wherefore, though some thought may be the ring-leader of others, yet [32] many are made without leaders.

Again, your author in his description of the mind says that "the discourse of the mind, when it is governed by design, is nothing but seeking, or the faculty of invention; a hunting out of the causes of some effects, present or past; or of the effects of some present or past cause [. . .]."[45] [. . .] In which discourse I do not perceive that he defines what the mind is, but I say that if, according to his opinion, nothing moves itself, but one thing moves another, then the mind must do nothing but move backward and forward, nay, only forward, and if all actions were thrusting or pressing of parts, it would be like a crowd of people, and there would be but little or no motion; for the crowd would make a stoppage, like water in a glass, the mouth of the glass being turned downward, no water can pass out, by reason the numerous drops are so closely pressed as they cannot move exteriously.

Next, I cannot conceive how the mind can run back [33] either to time or place, for as for place, the mind is enclosed in the body, and the running about in the parts of the body or brain will not inform it of an exterior place or object; besides, objects being the cause of the mind's motion, it must return to its cause, and so move until it come to the object that moved it first, so that the mind must run out of the body to that object, which moved it to such a thought, although that object were removed out of the world (as the phrase is). But for the mind to move backward to time past is more than it can do. Wherefore, in my opinion, remembrance or the like is only a repetition of such figures as were like to the objects; and for thoughts in particular, they are several figures made by the mind, which is the rational part of matter, in its own substance, either voluntarily or by imitation, whereof you may see more in my book of *Philosophical Opinions*.[46]

Hence I conclude that prudence is nothing else but a comparing of figures to figures, and of the several actions of those figures, as repeating former figures and comparing them to others of the like nature, qualities, proprieties, as also chances, fortunes, etc. Which figuring and repeating is done actually, in and by the rational matter, so that all the observation of the mind on outward objects is only an actual repetition of the mind, as moving in such or such figures and actions; and when the mind makes voluntary figures with those repeated figures and compares them together, this comparing is examination; and when several

---

45. Hobbes, *Leviathan*, part 1, chapter 3.

46. Cavendish begins discussing the human mind in part 2, chapter 9 of *Philosophical and Physical Opinions* (1663). Her account of how rational and sensitive matter form thoughts through "figuring" can be found in part 2, chapters 11–12.

figures agree and join, it is conclusion or judgment. Likewise does experience proceed from repeating and comparing of several figures in the mind, and [34] the more several figures are repeated and compared, the greater the experience is.

One thing more there is in the same chapter, which I cannot let pass without examination. Your author says that "things present only have a being in nature, things past only a being in the memory, but things to come have no being at all."[47] Which how it can possibly be, I am not able to conceive; for certainly, if nothing in nature is lost or annihilated, what is past and what is to come has as well a being as what is present; and, if that which is now, had its being before, why may it not also have its being hereafter? It might as well be said, that which is once forgot, cannot be remembered; for whatsoever is in nature has as much a being as the mind, and there is not any action, or motion, or figure, in nature, but may be repeated, that is, may return to its former figure when it is altered and dissolved. But by reason nature delights in variety, repetitions are not so frequently made, especially of those things or creatures which are composed by the sensitive corporeal motions in the inanimate part of matter, because they are not so easily wrought, as the rational matter can work upon its own parts, being more pliant in itself than the inanimate matter is. And this is the reason that there are so many repetitions of one and the same figure in the rational matter, which is the mind, but seldom any in the gross and inanimate part of matter, for nature loves ease and freedom.

But to conclude, Madam, I perceive your author confines sense only to animal-kind, and reason only to mankind. Truly, it is out of self-love, when one creature prefers his own excellency before another, for nature being endued with [35] self-love, all creatures have self-love, too, because they are all parts of nature, and when parts agree or disagree, it is out of interest and self-love; but man herein exceeds all the rest, as having a supernatural soul, whose actions also are supernatural. To which I leave him, and rest, Madam,

Your faithful friend and servant.

---

47. Hobbes, *Leviathan*, part 1, chapter 3.

9.

Madam,

When your author discourses of the use of "speech or words and names," he is pleased to say "that their use is to serve for marks and notes of remembrance,"[48] whereof to give you my opinion, I say that speech is natural to the shape of man; and though sometimes it serves for marks or notes of remembrance, yet it does not always, for all other animals have memory without the help of speech, and so have deaf and dumb men, nay more than those that hear and speak. Wherefore, though words are useful to the mind, and so to the memory, yet both can be without them, whereas words cannot be without memory; for take a bird and teach him to speak, if he had not memory before he heard the words, he could never learn them.

You will ask me, Madam, what then, [36] is memory the cause of speech? I answer, life and knowledge, which is sense and reason, as it creates and makes all sorts of creatures, so also amongst the rest it makes words. And as I said before that memory may be without the help of speech or words, so I say also that there is a possibility of reckoning of numbers, as also of magnitudes, of swiftness, of force, and other things without words, although your author denies it. But some men are so much for art as they endeavor to make art, which is only a drudgery-maid of nature, the chief mistress, and nature her servant, which is as much as to prefer effects before the cause, nature before God, discord before unity and concord.

Again, your author, in his chapter of reason, defines reason to be nothing else but reckoning.[49] I answer that in my opinion reckoning is not reason itself, but only an effect or action of reason; for reason, as it is the chiefest and purest degree of animate matter, works variously and in diverse motions, by which it produces various and diverse effects, which are several perceptions, as conception, imagination, fancy, memory, remembrance, understanding, judgment, knowledge, and all the passions, with many more. Wherefore this reason is not in one undivided part, nor bound to one motion, for it is in every creature more or less, and moves in its own parts variously; and in some creatures, as for example, in some men, it moves more variously than in others, which is the cause that some men are more dull and stupid than others; neither does reason always move in one creature regularly, which is the cause that some men are

---

48. Hobbes, *Leviathan*, part 1, chapter 4.
49. Hobbes, *Leviathan*, part 1, chapter 5.

mad or foolish. And though all men are made by the direction of [37] reason, and endued with reason from the first time of their birth, yet all have not the like capacities, understandings, imaginations, wits, fancies, passions, etc., but some more, some less, and some regular, some irregular, according to the motions of reason or rational part of animate matter; and though some rational parts may make use of other rational parts, as one man of another man's conceptions, yet all these parts cannot associate together; as for example, all the material parts of several objects, no not their species, cannot enter or touch the eye without danger of hurting or loosing it, nevertheless the eye makes use of the objects by patterning them out, and so does the rational matter, by taking patterns from the sensitive. And thus knowledge or perception of objects, both sensitive and rational, is taken without the pressure of any other parts; for though parts join to parts (for no part can be single), yet this joining does not necessarily infer[50] the pressure of objects upon the sensitive organs. Whereof I have already discoursed sufficiently heretofore, to which I refer you, and rest, Madam,

Your faithful friend and servant.

---

[38]   **10.**

Madam,

Understanding, says your author, "is nothing else but conception caused by speech, and therefore, if speech be peculiar to[51] man (as, for aught I know, it is), then is understanding peculiar to him also,"[52] where he confines understanding only to speech and to mankind. But, by his leave, Madam, I surely believe that there is more understanding in nature than that which is in speech, for if there were not, I cannot conceive how all the exact forms in generations[53] could be produced, or how there could be such distinct degrees of several sorts and kinds of creatures, or distinctions of times and seasons, and so many exact motions and figures in nature. Considering all this, my reason persuades me that all understanding, which is a part of knowledge, is not caused by speech, for all the

---

50. Infer: imply.
51. Peculiar to: uniquely belonging to.
52. Hobbes, *Leviathan*, part 1, chapter 4.
53. Generation: procreation, reproduction.

motions of the celestial orbs are not made by speech, neither is the knowledge or understanding which a man has when sick, as to know or understand he is sick, made by speech, nor by outward objects, especially in a disease he never heard, nor saw, nor smelt, nor tasted, nor touched. Wherefore all perception, sensation, memory, imagination, appetite, understanding, and the like, are not made nor caused by outward objects, nor by speech.

And as for the names of things, they are but different postures of the figures in our mind or thoughts, made by rational matter. But reasoning [39] is a comparing of the several figures with their several postures and actions in the mind, which, joined with the several words made by the sensitive motions, inform another distinct and separate part, as another man, of their mind's conceptions, understanding, opinions, and the like.

Concerning addition and subtraction, wherein your author says reasoning consists,[54] I grant that it is an act of reasoning, yet it does not make sense or reason, which is life and knowledge; but sense and reason, which is self-motion, makes addition and subtraction of several parts of matter; for had matter not self-motion, it could not divide nor compose, nor make such varieties, without great and lingering retardments, if not confusion.

Wherefore all what is made in nature is made by self-moving matter, which self-moving matter does not at all times move regularly, but often irregularly, which causes false logic, false arithmetic, and the like; and if there be not a certainty in these self-motions or actions of nature, much less [is there a certainty] in art, which is a secondary action; and therefore, neither speech, words, nor exterior objects cause understanding or reason. And although many parts of the rational and sensitive matter joined into one may be stronger by their association, and overpower other parts that are not so well knit and united, yet these are not the less pure; only the parts and motions being not equal in several creatures make their knowledge and reason more or less. For, when a man has more rational matter well regulated, and so more wisdom than another, that same man may chance to overpower the other, whose rational matter is more irregular [40], but yet not so much by strength of the united parts as by their subtlety; for the rational matter moving regularly is more strong with subtlety than the sensitive with force; so that wisdom is stronger than life, being more pure, and

---

54. Hobbes maintains that reasoning is analogous to mathematics; for example, inferring a general conclusion from particulars is analogous to addition (*Leviathan*, part 1, chapter 5).

so more active; for in my opinion, there is a degree of difference between life and knowledge, as my book of *Philosophical Opinions* will inform you.[55]

Again, your author says "that man doth excel all other animals in this faculty, that when he conceives anything whatsoever, he is apt to enquire the consequences of it, and what effects he can do with it. Besides this" (says he), "man has another degree of excellence, that he can by words reduce the consequences he finds to general rules called theorems or aphorisms, that is, he can reason or reckon not only by number, but in all other things, whereof one may be added unto, or subtracted from another."[56]

To which I answer that, according to my reason, I cannot perceive but that all creatures may do as much; but by reason they do it not after the same manner or way as man, man denies they can do it at all; which is very hard; for what man knows whether fish do not know more of the nature of water, and ebbing and flowing, and the saltiness of the sea; or whether birds do not know more of the nature of and degrees of air, or the cause of tempests; or whether worms do not know more of the nature of earth, and how plants are produced; or bees of the several sorts of juices of flowers, than men? And whether they do not make their aphorisms and theorems by their manner of intelligence? For though they have not the speech of man, yet thence does not [41] follow that they have no intelligence at all.

But the ignorance of men concerning other creatures is the cause of despising other creatures, imagining themselves as petty gods in nature, when as nature is not capable to make one God, much less so many as mankind; and were it not for man's supernatural soul, man would not be more supreme than other creatures in nature. "But," (says your author), "this privilege in man is allayed by another, which is, no living creature is subject to absurdity, but only man."[57] Certainly, Madam, I believe the contrary, to wit, that all other creatures do as often commit mistakes and absurdities as man, and if it were not to avoid tediousness, I could present sufficient proofs to you.

Wherefore I think not only man but also other creatures may be philosophers and subject to absurdities as aptly as men; for man does [not] nor cannot truly know the faculties and abilities or actions of all other creatures, no not of his own kind as mankind, for if he do measure all men by himself he will be very much mistaken, for what he conceives to be true or wise, another may conceive

---

55. See *Philosophical and Physical Opinions* (1663), part 1, chapters 15–16.

56. Hobbes, *Leviathan*, part 1, chapter 5.

57. Hobbes, *Leviathan*, part 1, chapter 5.

to be false and foolish. But man may have one way of knowledge in philosophy and other arts, and other creatures another way, and yet other creatures' manner or way may be as intelligible and instructive to each other as man's, I mean, in those things which are natural. Wherefore I cannot consent to what your author says, "that children are not endued with reason at all, till they have attained the use of speech";[58] for reason is in those creatures which have no speech, witness horses, especially those which are taught the manage,[59] and many other animals. And as for the [42] weak understanding in children, I have discoursed thereof in my book of philosophy;[60] the rest of this discourse, lest I tire you too much at once, I shall reserve for the next, retiring in the meantime, Madam,

Your faithful friend and servant.

11.

Madam,

I sent you word in my last that your author's opinion is "that children are not endued with reason at all, until they have attained to the use of speech"; in the same chapter he speaks to the same purpose thus: "Reason is not as sense and memory born with us, nor gotten by experience only, as prudence is, but attained by industry."[61]

To which I reply only this, that it might as well be said, a child when new-born has not flesh and blood, because by taking in nourishment or food, the child grows to have more flesh and blood; or that a child is not born with two legs, because he cannot go, or with arms and hands, because he cannot help himself; or that he is not born with a tongue, because he cannot speak. For although reason does not move in a child as in a man, in infancy as in youth, yet that does not prove that children are without reason, because they cannot run and prate.[62] [43] I grant, some other creatures appear to have more knowledge when new-born than others; as for example, a young foal has more knowledge than a young

---

58. Hobbes, *Leviathan*, part 1, chapter 5.
59. Manage: the art of training horses.
60. See *Philosophical and Physical Opinions* (1663), part 2, chapters 2–3.
61. Hobbes, *Leviathan*, part 1, chapter 5 (misidentified by Cavendish as chapter 4).
62. Prate: talk, chatter.

child, because a child cannot run and play; besides, a foal knows his own dam,[63] and can tell where to take his food, as to run and suck his dam, when as an infant cannot do so, nor all beasts, although most of them can, but yet this does not prove that a child has no reason at all.

Neither can I perceive that man is a monopoler of all reason, or animals of all sense, but that sense and reason are in other creatures as well as in man and animals; for example, drugs, as vegetables and minerals, although they cannot slice, pound, or infuse, as man can do, yet they can work upon man more subtly, wisely, and as sensibly, either by purging, vomiting, spitting, or any other way, as man by mincing, pounding, and infusing them, and vegetables will as wisely nourish men, as men can nourish vegetables. Also some vegetables are as malicious and mischievous to man, as man is to one another, witness hemlock, nightshade, and many more; and a little poppy will as soon, nay sooner cause a man to sleep, although silently, than a nurse a child with singing and rocking. But because they do not act in such manner or way as man, man judges them to be without sense and reason; and because they do not prate and talk as man, man believes they have not so much wit as he has; and because they cannot run and go, man thinks they are not industrious; the like for infants concerning reason.

But certainly, it is not local motion or speech that makes sense and reason, but sense and reason makes them; neither is sense and reason bound only to the actions of [44] man, but it is free to the actions, forms, figures, and proprieties of all creatures; for if none but man had reason, and none but animals sense, the world could not be so exact and so well in order as it is; but nature is wiser than man with all his arts, for these are only produced through the variety of nature's actions, and disputes [are only produced] through the superfluous varieties of man's follies and ignorances, not knowing nature's powerful life and knowledge. But I wonder, Madam, that your author says in this place "that reason is not born with man," when as in another place, he says "that every man brought philosophy, that is natural reason with him into the world."[64] Which how it agree, I will leave to others to judge, and to him to reconcile it, remaining in the meantime, Madam,

Your constant friend and faithful servant.

---

63. Dam: mother.

64. Thomas Hobbes, *Elements of Philosophy, the First Section, Concerning Body* (1656), part 1, chapter 1, section 1.

12.

Madam,

Two sorts of motions, I find your author does attribute to animals, *viz.*, "vital and animal." "The vital motions," says he, "are begun in generation, and continued without interruption through their whole life, and those are the course of the blood, the pulse, the breathing, [concoction],[65] nutrition, excretion, etc., to [45] which motions there needs no help of imaginations; but the animal motions otherwise called 'voluntary' motions are to go, to speak, to move any of our limbs, in such manner as is first fancied in our minds. And because going, speaking, and the like voluntary motions depend always upon a precedent thought of whither, which way, and what, it is evident that the imagination is the first internal beginning of all voluntary motion."[66] Thus far your author.

Whereof in short I give you my opinion, first concerning vital motions, that it appears improbable if not impossible to me that generation should be the cause and beginning of life, because life must of necessity be the cause of generation, life being the generator of all things, for without life motion could not be, and without motion not any thing could be begun, increased, perfected, or dissolved. Next, that imagination is not necessary to vital motions, it is probable it may not, but yet there is required knowledge, which I name "reason"; for if there were not knowledge in all generations or productions, there could not any distinct creature be made or produced, for then all generations would be confusedly mixed, neither would there be any distinct kinds or sorts of creatures, nor no different faculties, proprieties, and the like.

Thirdly, concerning animal motions, which your author names "voluntary motions," "as to go, to speak, to move any of our limbs, in such manner as is first fancied in our minds, and that they depend upon a precedent thought of whither, which way, and what, and that imagination is the first internal beginning of them"; I think, by your author's leave, it does imply a contradiction to call them "voluntary" motions, and yet to say they are caused and depend upon our [46] imagination; for if the imagination draws them this way or that way, how can they be voluntary motions, being in a manner forced and necessitated to move according to fancy or imagination?

---

65. Concoction: digestion. Cavendish mistakenly wrote "conviction" here.
66. Hobbes, *Leviathan*, part 1, chapter 6.

But when he goes on in the same place and treats of endeavor, appetite, desire, hunger, thirst, aversion, love, hate, and the like, he derives one from the other, and treats well as a moral philosopher; but whether it be according to the truth or probability of natural philosophy, I will leave to others to judge, for in my opinion passions and appetites are very different, appetites being made by the motions of the sensitive life, and passions, as also imagination, memory, etc. by the motions of the rational life, which is the cause that appetites belong more to the actions of the body than the mind. 'Tis true, the sensitive and rational self-moving matter does so much resemble each other in their actions as it is difficult to distinguish them. But having treated hereof at large in my other philosophical work,[67] to cut off repetitions, I will refer you to that, and desire you to compare our opinions together. But certainly there is so much variety in one and the same sort of passions, and so of appetites, as it cannot be easily expressed.

To conclude, I do not perceive that your author tells or expresses what the cause is of such or such actions, only he mentions their dependence, which is as if a man should converse with a nobleman's friend or servant, and not know the lord himself. But leaving him for this time, it is sufficient to me that I know your Ladyship, and your Ladyship knows me, that I am, Madam,

Your faithful friend and humble servant.

[47]   **13.**

Madam,

Having obeyed your commands in giving you my opinion of the first part of the book of that famous and learned author you sent me, I would go on; but seeing he treats in his following parts of the politics, I was forced to stay my pen, because of these following reasons.

First, that a woman is not employed in state affairs, unless an absolute queen. Next, that to study the politics is but loss of time, unless a man were sure to be a favorite to an absolute prince. Thirdly, that it is but a deceiving profession, and requires more craft than wisdom. All which considered, I did not read that part of your author. But as for his natural philosophy, I will send you my opinion so

---

67. On passions, appetites, and the relationship between sensitive and rational matter, see *Philosophical and Physical Opinions* (1663), part 6, chapter 9.

far as I understand it. For what belongs to art, as to geometry, being no scholar, I shall not trouble myself withal.

And so I'll take my leave of you when I have in two or three words answered the question you sent me last, which was whether nature be the art of God, man the art of nature, and a politic[68] government the art of man? To which I answer, 'tis probable it may be so; only I add this, that nature does not rule God, nor man nature, nor politic government man. For the effect cannot rule the cause, but the cause does rule the effect. Wherefore if men do not naturally agree, art cannot make unity amongst them, or associate them into one politic body and so [48] rule them. But man thinks he governs, when as it is nature that does it, for as nature does unite or divide parts regularly or irregularly, and moves the several minds of men and the several parts of men's bodies, so war is made or peace kept. Thus it is not the artificial form that governs men in a politic government, but a natural power, for though natural motions can make artificial things, yet artificial things cannot make natural power; and we might as well say nature is governed by the art of nature as to say man is ruled by the art and invention of man.

The truth is, man rules an artificial government, and not the government man, just like as a watchmaker rules his watch, and not the watch the watchmaker. And thus I conclude and rest, Madam,

Your faithful friend and servant.

[52] **16.**

Madam,

"An accident," says your author, "is nothing else, but the manner of our conception of body, or that faculty of any body, by which it works in us a conception of itself";[69] to which I willingly consent; but yet I say that these qualities cannot be separated from the body, for as impossible [as] it is that the essence of nature should be separable from nature, [just] as impossible is it that the various modes or alterations, either of figures or motions, should be separable from matter or body. Wherefore when he goes on and says, "an accident is not a body, but in a body, yet not so as if anything were contained therein, as if for example, redness

68. Politic: political.
69. Hobbes, *Elements*, part 2, chapter 8, section 2.

were in blood in the same manner as blood is in a bloody cloth; but as magnitude is in that which is great, rest in that which rests, motion in that which is moved";[70] I answer that in my opinion, not any thing in nature can be without a body, and that redness is as well in blood, as blood is in a bloody cloth, or any other color in anything else; for there is no color without a body, but every color has as well a body as anything else, and if color be a separable accident, I would fain know how it can be separated from a subject, being bodiless, for that which is no body is nothing, and nothing cannot be taken away from anything. Wherefore as for natural color, it cannot be taken away from any creature without the parts of its substance or body; and as for artificial [53] colors, when they are taken away, it is a separation of two bodies which joined together; and if color, or hardness, or softness do change, it is nothing else but an alteration of motions and not an annihilation, for all changes and alterations remain in the power of corporeal motions, as I have said in other places;[71] for we might as well say life does not remain in nature, when a body turns from an animal to some other figure, as believe that those they name "accidents" do not remain in corporeal motions.

Wherefore I am not of your author's mind when he says that "when a white thing is made black, the whiteness perishes";[72] for it cannot perish, although it is altered from white to black, being in the power of the same matter to turn it again from black to white, so as it may make infinite repetitions of the same thing; but by reason nature takes delight in variety, she seldom uses such repetitions; nevertheless that does not take away the power of self-moving matter, for "it does not" and "it cannot" are two several things, and the latter does not necessarily follow upon the former.

Wherefore not any the least thing can perish in nature, for if this were possible, the whole body of nature might perish also, for if so many figures and creatures should be annihilated and perish without any supply or new creation, nature would grow less and at last become nothing; besides, it is as difficult for nature to turn something into nothing as to create something out of nothing. Wherefore as there is no annihilation or perishing in nature, so there is neither any new creation in nature.

But your author makes a difference between bodies and accidents, saying "that bodies are things and not generated, but accidents are generated and

---

70. Hobbes, *Elements*, part 2, chapter 8, section 3.

71. See, for example, *Philosophical and Physical Opinions* (1663), part 4, chapter 3.

72. Hobbes, *Elements*, part 2, chapter 8, section 19 (misidentified by Cavendish as section 20).

not [54] things."[73] Truly, Madam, these accidents seem to me to be like van Helmont's lights, gases, blazes, and ideas, and Dr. More's immaterial substances or demons; only in this Dr. More has the better, that his immaterial substances are beings which subsist of themselves, whereas accidents do not, but their existence is in other bodies. But what they call "accidents" are in my opinion nothing else but corporeal motions, and if these accidents be generated, they must needs be bodies, for how nothing can be generated in nature is not conceivable, and yet your author denies that "accidents are something, namely some part of a natural thing."[74] But as for generations, they are only various actions of self-moving matter, or a variety of corporeal motions, and so are all accidents whatsoever, so that there is not any thing in nature that can be made new or destroyed, for whatsoever was and shall be, is in nature, though not always in act, yet in power, as in the nature and power of corporeal motions, which is self-moving matter.

And as there is no new generation of accidents, so there is neither a new generation of motions; wherefore when your author says that "when the hand, being moved, moves the pen, the motion does not go out of the hand into the pen, for so the writing might be continued, though the hand stood still, but a new motion is generated in the pen, and is the pen's motion,"[75] I am of his opinion that the motion does not go out of the hand into the pen, and that the motion of the pen is the pen's own motion; but I deny that after holding the hand a little while still and beginning to write again, a new motion of the pen is generated; for it is only a repetition, and not a new generation, for the hand, pen, [55] and ink repeat but the same motion or action of writing. Besides, generation is made by connection or conjunction of parts, moving by consent to such or such figures, but the motion of the hand or the pen is always one and the same; wherefore it is but the variation and repetition in and of the same motion of the hand or pen, which may be continued in that manner infinitely, just as the same corporeal motions can make infinite variations and repetitions of one and the same figure, repeating it as oft as they please, as also making copy of copy.

And although I do not deny but there are generations in nature, yet not annihilations or perishings, for if any one motion or figure should perish, the matter must perish also; and if any one part of matter can perish, all the matter in nature may perish also; and if there can any new thing be made or created in nature which has not been before, there may also be a new nature; and so by

---

73. Hobbes, *Elements*, part 2, chapter 8, section 19.

74. Hobbes, *Elements*, part 2, chapter 8, section 2.

75. Hobbes, *Elements*, part 2, chapter 8, section 21.

perishings and new creations, this world would not have continued an age. But surely whatsoever is in nature has been existent always.

Wherefore to conclude, it is not the generation and perishing of an accident that makes its subject to be changed, but the production and alteration of the form makes it said to be generated or destroyed, for matter will change its motions and figures without perishing or annihilating; and whether there were words or not, there would be such causes and effects. But having not the art of logic to dispute with artificial words, nor the art of geometry to demonstrate my opinions by mathematical figures, I fear they will not be so [56] well received by the learned. However, I leave them to any man's unprejudiced reason and judgment, and devote myself to your service, as becomes, Madam,

Your Ladyship's humble and faithful servant.

17.

Madam,

Your author, concerning place and magnitude, says that "place is nothing out of the mind, nor magnitude anything within it; for place is a mere phantasm of a body of such quantity and figure; and magnitude a peculiar accident of the body."[76] But this does not well agree with my reason, for I believe that place, magnitude, and body are but one thing, and that place is as true an extension as magnitude, and not a feigned one.

Neither am I of his opinion "that place is immoveable,"[77] but that place moves according as the body moves, for not any body wants place, because place and body is but one thing, and wheresoever is body, there is also place, and wheresoever is place, there is body, as being one and the same. Wherefore motion cannot be "a relinquishing of one place and acquiring another,"[78] for there is no such thing as place different from body, but what is called "change of place" is nothing [57] but change of corporeal motions; for say a house stands in such a place, if the house be gone, the place is gone also, as being impossible that the place of the house should remain when the house is taken away; like as a man

---

76. Hobbes, *Elements*, part 2, chapter 8, section 5.

77. Hobbes, *Elements*, part 2, chapter 8, section 5.

78. Hobbes, *Elements*, part 2, chapter 8, section 10.

when he is gone out of his chamber, his place is gone too. 'Tis true, if the ground or foundation do yet remain, one may say there stood such a house heretofore, but yet the place of the house is not there really at that present, unless the same house be built up again as it was before, and then it has its place as before. Nevertheless the house being not there, it cannot be said that either place or house are annihilated, *viz.*, when the materials are dissolved, no not when transformed into millions of several other figures, for the house remains still in the power of all those several parts of matter; and as for "space," it is only a distance between some parts or bodies. But an "empty place" signifies to my opinion nothing; for if place and body are one and the same, and empty is as much as nothing, then certainly these two words cannot consist together, but are destructive to one another.

Concerning that your author says, "two bodies cannot be in the same place, nor one body in two places at the same time,"[79] is very true, for there are no more places than bodies, nor more bodies than places, and this is to be understood as well of the grosser as the purest parts of nature, of the mind as well as of the body, of the rational and sensitive animate matter as well as of the inanimate, for there is no matter, how pure and subtle soever, but is embodied, and all that has body has place.

Likewise, I am of his opinion "that one body has always one and the same magnitude";[80] for, in my opinion, magnitude, place and [58] body do not differ, and as place, so magnitude can never be separated from body. But when he speaks of "rest," I cannot believe there is any such thing truly in nature, for it is impossible to prove that anything is without motion, either consistent, or composing, or dissolving, or transforming motions, or the like, although not altogether perceptible by our senses, for all the matter is either moving or moved, and although the moved parts are not capable to receive the nature of self-motion from the self-moving parts, yet these self-moving parts, being joined and mixed with all the other parts of the moved matter, do always move the same; for the moved or inanimate part of matter, although it is a part of itself, yet it is so intermixed with the self-moving animate matter as they make but one body; and though some parts of the inanimate may be as pure as the sensitive animate matter, yet they are never so subtle as to be self-moving. Wherefore the sensitive moves in the inanimate, and the rational in the sensitive, but often the rational moves in itself. And although there is no rest in nature, nevertheless matter

---

79. Hobbes, *Elements*, part 2, chapter 8, section 8.
80. Hobbes, *Elements*, part 2, chapter 8, section 5.

could have been without motion, when as it is impossible that matter could be without place or magnitude, no more than variety can be without motion. And thus much at this present. I conclude, and rest, Madam,

Your faithful friend and servant.

[59]   **18.**

Madam,

Passing by those chapters of your author that treat of power and act, identity and difference, analogism, angle and figure, figures deficient, dimension of circles, and several others, most of which belong to art, as geometry and the like, I am come to that wherein he discourses of sense and animal motion, saying "that some natural bodies have in themselves the patterns almost of all things, and others of none at all."[81] Whereof my opinion is that the sensitive and rational parts of matter are the living and knowing parts of nature, and no part of nature can challenge them only to itself, nor no creature can be sure that sense is only in animal-kind and reason in mankind; for can anyone think or believe that nature is ignorant and dead in all her other parts besides animals? Truly this is a very unreasonable opinion; for no man, as wise as he thinks himself, nay were all mankind joined into one body, yet they are not able to know it, unless there were no variety of parts in nature, but only one whole and individable body, for other creatures may know and perceive as much as animals, although they have not the same sensitive organs, nor the same manner or way of perception.

Next your author says, "the cause of sense or perception consists herein, that the first organ of sense is touched and pressed; for when the uttermost part of the organ is pressed, it no sooner yields, [60] but the part next within it is pressed also, and in this manner the pressure or motion is propagated through all the parts of the organ to the innermost. And thus also the pressure of the uttermost part proceeds from the pressure of some more remote body, and so continually, till we come to that, from which, as from its fountain, we derive the phantasm or idea, that is made in us by our sense. And this, whatsoever it be, is that we commonly call the 'object'; sense therefore is some internal motion in the sentient, generated by some internal motion of the parts of the object, and

---

81. Hobbes, *Elements*, part 4, chapter 25, section 1.

propagated through all the media to the innermost part of the organ. Moreover there being a resistance or reaction in the organ, by reason of its internal motion against the motion propagated from the object, there is also an endeavor in the organ opposite to the endeavor proceeding from the object, and when that endeavor inwards is the last action in the act of sense, then from the reaction a phantasm or idea has its being."[82]

This is your author's opinion, which if it were so, perception could not be effected so suddenly, nay I think the sentient by so many pressures in so many perceptions would at last be pressed to death; besides, the organs would take a great deal of hurt, nay totally be removed out of their places, so as the eye would in time be pressed into the center of the brain. And if there were any resistance, reaction, or endeavor in the organ, opposite to the endeavor of the object, there would, in my opinion, be always a war between the animal senses and the objects, the endeavor of the objects pressing one way and the senses pressing the other way, and if equal in their strengths, they would make a stop, and the sensitive organs would be very much pained. Truly, Madam, in [61] my opinion it would be like that custom which formerly has been used at Newcastle, when a man was married, the guests divided themselves, behind and before the bridegroom, the one party driving him back, the other forward, so that one time a bridegroom was killed in this fashion.

But certainly nature has a more quick and easy way of giving intelligence and knowledge to her creatures, and does not use such constraint and force in her actions. Neither is sense or sensitive perception a mere phantasm or idea, but a corporeal action of the sensitive and rational matter; and according to the variation of the objects or patterns, and the sensitive and rational motions, the perception also is various, produced not by external pressure, but by internal self-motion, as I have declared heretofore; and to prove that the sensitive and rational corporeal motions are the only cause of perception, I say, if those motions in an animal move in another way, and not to such perceptions, then that animal can neither hear, see, taste, smell, nor touch, although all his sensitive organs be perfect, as is evident in a man falling into a swoon, where all the time he is in a swoon, the pressure of the objects is made without any effect. Wherefore, as the sensitive and rational corporeal motions make all that is in nature, so likewise they make perception, as being perception itself, for all self-motion is perception, but all perception is not animal perception, or even

---

82. Hobbes, *Elements*, part 4, chapter 25, section 2.

after an animal way; and therefore sense cannot decay nor die, but what is called a "decay" or "death" is nothing else but a change or alteration of those motions.

But you will say, Madam, it may be that one body, as an object, leaves the print of its figure in the next [62] adjoining body, until it comes to the organ of sense. I answer then that soft bodies only must be pressed, and the object must be so hard as to make a print, and as for rare[83] parts of matter, they are not able to retain a print without self-motion. Wherefore it is not probable that the parts of air should receive a print, and print the same again upon the adjoining part, until the last part of the air print it upon the eye; and that the exterior parts of the organ should print upon the interior, till it come to the center of the brain, without self-motion.

Wherefore in my opinion, perception is not caused either by the printing of objects, nor by pressures, for pressures would make a general stop of all natural motions, especially if there were any reaction or resistance of sense; but according to my reason, the sensitive and rational corporeal motions in one body pattern out the figure of another body, as of an exterior object, which may be done easily without any pressure or reaction. I will not say that there is no pressure or reaction in nature, but pressure and reaction do not make perception, for the sensitive and rational parts of matter make all perception and variety of motion, being the most subtle parts of nature, as self-moving, as also dividable, and composable, and alterable in their figurative motions, for this perceptive matter can change its substance into any figure whatsoever in nature, as being not bound to one constant shape. But having treated hereof before, and being [resolved] to say more of it hereafter, this shall suffice for the present, remaining always, Madam,

Your constant friend and faithful servant.

[63]   **19.**

Madam,

To discourse of the world and stars is more than I am able to do, wanting the art of astronomy and geometry; wherefore, passing by that chapter of your author, I am come to that wherein he treats of light, heat, and colors;[84] and to

---

83. Rare: rarefied, not densely packed.
84. Hobbes, *Elements*, part 4, chapter 27.

give you my opinion of light, I say it is not the light of the sun that makes an animal see, for we can see inwardly in dreams without the sun's light, but it is the sensitive and rational motions in the eye and brain that make such a figure as light. For if light did press upon the eye, according to your author's opinion, it might put the eye into as much pain as fire does, when it sticks its points into our skin or flesh. The same may be said of colors, for the sensitive motions make such a figure, which is such a color, and such a figure, which is such a color.

Wherefore light, heat, and color, are not bare and bodiless qualities, but such figures made by corporeal self-motions, and are as well real and corporeal objects as other figures are; and when these figures change or alter, it is only that their motions alter, which may alter and change heat into cold, and light into darkness, and black color into white.

But by reason the motions of the sun are so constant as the motions of any other kind of creatures, it is no more subject to be altered than all the world, unless nature did it by the command of God; for though the parts [64] of self-moving matter be alterable, yet all are not altered; and this is the reason that the figure of light in our eye and brain is altered, as well as it is alterable, but not the real figure of the sun, neither does the sun enter our eyes; and as the light of the sun is made or patterned in the eye, so is the light of glow-worms' tails and cats' eyes that shine in the dark, made not by the sun's but their own motions in their own parts. The like when we dream of light, the sensitive corporeal motions working inwardly make the figure of light on the inside of the eye, as they did pattern out the figure of light on the outside of the eye when awake, and the objects before them; for the sensitive motions of the eye pattern out the figure of the object in the eye, and the rational motions make the same figure in their own substance.

But there is some difference between those figures that perceive light and those that are light themselves; for when we sleep, there is made the figure of light, but not from a copy; but when the eye sees light, that figure is made from a copy of the real figure of the sun; but those lights which are inherent, as in glow-worms' tails, are original lights, in which is as much difference as between a man and his picture; and as for the swiftness of the motions of light and the violence of the motions of fire, it is very probable they are so, but they are a certain particular kind or sort of swift and violent motions; neither will all sorts of swift and violent motions make fire or light, as for example the swift and violent circular motion of a whirlwind neither makes light nor fire. Neither is all fire light, nor all light fire, for there is a sort of dead fire, as in spices, spirits, oils, and the like;[85] and several sorts of lights, which are not hot, as [65] the light which is made in dreams, as also the inherent lights in glow-worms, cats' eyes, fish-bones, and the like; all which several fires and lights are made by the self-moving matter and motions distinguishable by their figures, for those motions make such a figure for the sun's light, such a figure for glow-worms' light, such a figure for cats' eyes' light, and so some alteration in every sort of light. The same for fire, only firelight is a mixed figure, as partly of the figure of fire and partly of the figure of light. Also colors are made after the like manner, *viz.* so many several colors, so many several figures; and as these figures are less or more different, so are the colors.

Thus, Madam, whosoever will study nature must consider the figures of every creature, as well as their motions, and must not make abstractions of motion and figure from matter, nor of matter from motion and figure, for they are inseparable as being but one thing, *viz.* corporeal figurative motions; and whosoever conceives any of them as abstract, will, in my opinion, very much err; but men are apt to make more difficulties and enforcements[86] in nature than nature ever knew.

But to return to light. There is no better argument to prove that all objects of sight are figured in the eye by the sensitive, voluntary, or self-motions, without the pressure of objects, but that not only the pressure of light would hurt the tender eye, but that the eye does not see all objects according to their magnitude, but sometimes bigger, sometimes less. As for example, when the eye looks through a small passage, as a perspective-glass,[87] by reason of the difficulty of seeing a body through a small hole, and the double figure of the glass [66] being convex and concave, the corporeal motions use more force, by which the object is enlarged, like as a spark of fire by force is dilated into a great fire, and a drop of water by blowing into a bubble; so the corporeal motions do double and treble their strength, making the image of the object exceeding large in the eye; for though the eye be contracted, yet the image in the eye is enlarged to a great extension; for the sensitive and rational matter is extremely subtle, by reason it is extremely pure, by which it has more means and ways of magnifying than the

---

85. In *Philosophical and Physical Opinions* (1663), Cavendish characterizes "dead fire" as "interiously hot and burning, yet not exteriously hot and burning" (part 5, chapter 11). That is, the interior motions of something with "dead fire" resemble those of something that is visibly burning, but the exterior motions do not.

86. Enforcements: acts of violence.

87. Perspective-glass: telescope.

perspective-glass. But I intend to write more of this subject in my next, and so I break off here, resting, Madam,

Your faithful friend and servant.

<hr>

20.

<hr>

Madam,

Some will perhaps question the truth or probability of my saying that light is a body, objecting that if light were a body, when the sun is absent or retires under our horizon, its light would leave an empty place, or if there were no empty place but all full, the light of the sun at its return would not have room to display itself, especially in so great a compass as it does, for two bodies cannot be in one place at one time.

I answer, [67] all bodies carry their places along with them, for body and place go together and are inseparable, and when the light of the sun is gone, darkness succeeds, and when darkness is gone, light succeeds, so that it is with light and darkness as with all creatures else; for you cannot believe that if the whole world were removed, there would be a place of the world left, for there cannot be an empty nothing, no more than there can be an empty something; but if the world were annihilated, the place would be annihilated too, place and body being one and the same thing; and therefore in my opinion, there be no more places than there are bodies, nor no more bodies than there are places.

Secondly, they will think it absurd that I say the eye can see without light; but in my opinion it seems not absurd, but very rational, for we may see in dreams, and some do see in the dark, not in their fancy or imagination, but really. And as for dreams, the sensitive corporeal motions make a light on the inside of the organ of sight really, as I have declared in my former letter. But that we do not see ordinarily without exterior light, the reason is that the sensitive motions cannot find the outward objects to pattern out without exterior light; but all perception does not proceed from light, for all other perception besides animal sight requires not light. Neither, in my opinion, does the perception of sight in all creatures but animals, but yet animals do often see in the dark and in sleep.

I will not say but that the animate matter which by self-motion does make the perception of light with other perceptive figures, and so animal perceptive light, may be the presenter or ground perceptive figure of sight; yet the sensitive corporeal [68] motions can make other figures without the help of light, and such as light did never present. But when the eye patterns out an exterior object presented by light, it patterns also out the object of light; for the sensitive motions can make many figures by one act, not only in several organs, but in one organ; as for example, there is presented to sight a piece of embroidery, wherein is silk, silver, and gold upon satin in several forms or figures, as several flowers, the sensitive motions straight by one and the same act pattern out all those several figures of flowers, as also the figures of silk, silver, gold, and satin, without any pressure of these objects or motions in the medium; for if they all should press, the eye would no more see the exterior objects than the nose, being stopped, could smell a presented perfume.

Thirdly, they may ask me, if sight be made in the eye and proceeds not from the outward object, what is the reason that we do not see inwardly, but outwardly, as from us? I answer, when we see objects outwardly, as from us, then the sensitive motions work on the outside of the organ, which organ, being outwardly convex, causes us to see outwardly, as from us, but in dreams we see inwardly; also the sensitive motions do pattern out the distance together with the object.

But you will say, the body of the distance, as the air, cannot be perceived, and yet we can perceive the distance. I answer, you could not perceive the distance but by such or such an object as is subject to your sight; for you do not see the distance more than the air, or the like rare body that is between grosser objects; for if there were no stars, no planets, nor clouds, nor earth, nor water, but only air, you would not see any space or [69] distance; but light being a more visible body than air, you might figure the body of air by light, but so as in an extensive or dilating way; for when the mind or the rational matter conceives anything that has not such an exact figure, or is not so perceptible by our senses, then the mind uses art, and makes such figures which stand like to that; as for example, to express infinite to itself, it dilates its parts without alteration, and without limitation or circumference. Likewise, when it will conceive a constant succession of time, it draws out its parts into the figure of a line; and if eternity, it figures a line without beginning and end. But as for immaterial, no mind can conceive that, for it cannot put itself into nothing, although it can dilate and rarify itself to a higher degree, but must stay within the circle of natural bodies, as I within the circle of your commands, to express myself, Madam,

Your faithful friend and obedient servant.

22.

Madam,

The generation of sound, according to your worthy author's opinion, is as follows: "As vision," says he, "so hearing is generated by the medium, but not in the same manner; for sight is from pressure, that is, from an endeavor, in which there is no perceptible progression of any of the parts of the medium, but one part urging or thrusting another propagates that action successively to any distance whatsoever; whereas the motion of the medium by which sound is made, is a stroke; for when we hear, the drum of the ear, which is the first organ of hearing, is stricken, and the drum being stricken, the pia mater[88] is also shaken, and with it the arteries inserted into it, by which the action propagated to the heart itself, by the reaction of the heart a phantasm is made which we call sound."[89] Thus far your author.

To which give me leave to reply, that I fear, if the ear was bound to hear any loud music or another sound a good while, it would soundly be beaten, and grow sore and bruised with so many strokes; but since a pleasant sound would be rendered very unpleasant in this manner, my opinion is that like as in the eye, so in the ear the corporeal sensitive motions do pattern out as many several figures as sounds are presented to them; but if these motions be irregular, then the figure of the sound in the ear is not perfect according to the original; for if it be that the motions are tired with figuring, or the object of sound [73] be too far distant from the sensitive organ, then they move slowly and weakly, not that they are tired or weak in strength, but with working and repeating one and the same object, and so through love to variety, change from working regularly to move irregularly, so as not to pattern outward objects as they ought, and then there are no such patterns made at all, which we call "to be deaf"; and sometimes the sensitive motions do not so readily perceive a soft sound near as a stronger farther off.

But to prove it is not the outward object of sound with its striking or pressing motion, nor the medium, that causes this perception of sense, if there be a great solid body, as a wall or any other partition between two rooms, parting the object and the sensitive organ so as the sound is not able to press it, nevertheless the perception will be made. And as for pipes to convey sounds, the perception is more fixed and perfecter in united than in dilated or extended bodies,

---

88. Pia mater: brain.
89. Hobbes, *Elements*, part 4, chapter 29, section 1.

and then the sensitive motions can make perfecter patterns; for the stronger the objects are, the more perfect are the figures and patterns of the objects, and the more perfect is the perception. But when the sound is quite out of the ear, then the sensitive motions have altered the patterning of such figures to some other action; and when the sound fades by degrees, then the figure or pattern alters by degrees; but for the most part the sensitive corporeal motions alter according as the objects are presented, or the perception patterns out. Neither do they usually make figures of outward objects, if not perceived by the sense, unless through irregularities, as in madmen, which see such and such things, when as these things are not near, and then the sensitive motions work by rote, or after [74] their own voluntary invention.

As for reflection, it is a double perception, and so a double figure of one object; like as many pictures of one man, where some are more perfect than others, for a copy of a copy is not so perfect as a copy of an original. But the recoiling of sound is that the sensitive motions in the ear begin a new pattern before they dissolved the former, so as there is no perfect alteration or change from making to dissolving, but pattern is made upon pattern, which causes a confusion of figures, the one being neither perfectly finished, nor the other perfectly made. But it is to be observed that not always [do] the sensitive motions in the organs take their pattern from the original, but from copies; as for example, the sensitive motions in the eye pattern out the figure of an eye in a glass,[90] and so do not take a pattern from the original itself, but by another pattern, representing the figure of the eye in a looking-glass.

The same does the ear, by patterning out echoes, which is but a pattern of a pattern. But when as a man hears himself speak or make a sound, then the corporeal sensitive motions in the ear pattern out the object or figure made by the motions of the tongue and the throat, which is voice; by which we may observe, that there may be many figures made by several motions from one original; as for example, the figure of a word is made in a man's mouth, then the copy of that figure is made in the ear, then in the brain, and then in the memory, and all this in one man. Also, a word being made in a man's mouth, the air takes a copy or many copies thereof; but the ear patterns them both out, first the original coming from the mouth, and then the copy made in the air, which is called an "echo," [75] and yet not any strikes or touches each other's parts, only perceives and patterns out each other's figure.

---

90. Glass: mirror.

Neither are their substances the same, although the figures be alike; for the figure of a man may be carved in wood, then cut in brass, then in stone, and so forth, where the figure may be always the same, although the substances which do pattern out the figure are several, *viz.* wood, brass, stone, etc., and so likewise may the figure of a stone be figured in the fleshy substance of the eye, or the figure of light or color, and yet the substance of the eye remains still the same; neither does the substantial figure of a stone or tree, patterned out by the sensitive corporeal motions in the flesh of an animal eye, change from being a vegetable or mineral to an animal, and if this cannot be done by nature, much less by art; for if the figure of an animal be carved in wood or stone, it does not give the wood or stone any animal knowledge, nor an animal substance, as flesh, blood, bones, etc. No more does the patterning or figuring of a tree give a vegetable knowledge, or the substance of wood to the eye, for the figure of an outward object does not alter the substance that patterns it out or figures it, but the patterning substance does pattern out the figure in itself, or in its own substance, so as the figure which is patterned has the same life and knowledge with the substance by and in which it is figured or patterned, and the inherent motions of the same substance. And according as the sensitive and rational self-moving matter moves, so figures are made; and thus we see that lives, knowledge, motions, and figures are all material, and all creatures are endued with life, knowledge, motion, and figure, but not all alike or after the same manner.

But [76] to conclude this discourse of perception of sound, the ear may take the object of sound afar off as well as at a near distance; not only if many figures of the same sound be made from that great distance, but if the interposing parts be not so thick, close, or many as to hinder or obscure the object from the animal perception in the sensitive organ; for if a man lays his ear near to the ground, the ear may hear at a far distance as well as the eye can see, for it may hear the noise of a troop afar off, perception being very subtle and active. Also, there may several copies be made from the original, and from the last copy nearest to the ear, the ear may take a pattern, and so pattern out the noise in the organ without any strokes to the ear, for the subtle matter in all creatures does inform and perceive.

But this is well to be observed, that the figures of objects are as soon made as perceived by the sensitive motions in their work of patterning. And this is my opinion concerning the perception of sound, which together with the rest I leave to your Ladyship's and others' wiser judgment, and rest, Madam,

Your faithful friend and servant.

Madam,

I perceive by your last that you cannot well apprehend my meaning when I say that the print or figure of a body printed or carved is not made by the motions of the body printing or carving it, but by the motions of the body or substance printed or carved; for say you, does a piece of wood carve itself, or a black patch of a lady[91] cut its own figure by its own motions?

Before I answer you, Madam, give me leave to ask you this question, whether it be the motion of the hand, or the instrument, or both, that print or carve such or such a body? Perchance you will say that the motion of the hand moves the instrument, and the instrument moves the wood which is to be carved. Then I ask, whether the motion that moves the instrument, be the instrument's or the hand's? Perchance you will say the hand's; but I answer, how can it be the hand's motion, if it be in the instrument?

You will say, perhaps, the motion of the hand is transferred out of the hand into the instrument, and so from the instrument into the carved figure; but give me leave to ask you, was this motion of the hand that was transferred corporeal or incorporeal? If you say corporeal, then the hand must become less and weak, but if incorporeal, I ask you, how a bodiless motion can have force and strength to carve and cut?

But put an impossible proposition, as that there is an immaterial motion, and [78] that this incorporeal motion could be transferred out of one body into another; then I ask you, when the hand and instrument cease to move, what is become of the motion? Perhaps you will say the motion perishes or is annihilated, and when the hand and the instrument do move again, to the carving or cutting of the figure, then a new incorporeal motion is created. Truly then there will be a perpetual creation and annihilation of incorporeal motions, that is, of that which naturally is nothing; for an incorporeal being is as much as a natural nothing, for natural reason cannot know nor have naturally any perception or idea of an incorporeal being. Besides, if the motion be incorporeal, then it must needs be a supernatural spirit, for there is not anything else immaterial but they, and then it will be either an angel or a devil, or the immortal soul of man. But if you say it is the supernatural soul, truly I cannot be persuaded that the supernatural soul should not have any other employment than to carve or cut prints

---

91. In early modern Europe, men and women sometimes used small patches of black fabric, often cut into shapes such as crescents or stars, to cover blemishes.

or figures, or move in the hands, or heels, or legs, or arms of a man; for other animals have the same kind of motions, and then they might have a supernatural soul as well as man, which moves in them.

But if you say, that these transferrable motions are material, then every action whereby the hand moves to the making or moving of some other body would lessen the number of the motions in the hand and weaken it, so that in the writing of one letter, the hand would not be able to write a second letter, at least not a third.

But I pray, Madam, consider rationally that though the artificer or workman be the occasion of the motions of the carved body, yet the motions of [79] the body that is carved are they which put themselves into such or such a figure, or give themselves such or such a print as the artificer intended; for a watch, although the artist or watchmaker be the occasional cause that the watch moves in such or such an artificial figure, as the figure of a watch, yet it is the watch's own motion by which it moves; for when you carry the watch about you, certainly the watchmaker's hand is not then with it as to move it; or if the motion of the watchmaker's hand be transferred into the watch, then certainly the watchmaker cannot make another watch, unless there be a new creation of new motions made in his hands; so that God and nature would be as much troubled and concerned in the making of watches as in the making of a new world. For God created this world in six days and rested the seventh day, but this would be a perpetual creation.

Wherefore I say that some things may be occasional causes of other things, but not the prime or principal causes; and this distinction is very well to be considered, for there are not frequenter mistakes than to confound these two different causes, which make so many confusions in natural philosophy, and this is the opinion of, Madam,

Your faithful friend and servant.

[80] **24.**

Madam,

In answer to your question, what makes echo, I say it is that which makes all the effects of nature, *viz.* self-moving matter. I know the common opinion is that echo is made like as the figure of a face or the like in a looking-glass, and that the reverberation of sound is like the reflection of sight in a looking-glass. But I am not of that opinion, for both echo and that which is called the reflection in a looking-glass are made by the self-moving matter, by way of patterning and copying out.

But then you will ask me, whether the glass takes the copy of the face, or the face prints its copy on the glass, or whether it be the medium of light and air that makes it? I answer, although many learned men say that as all perception, so also the seeing of one's face in a looking-glass, and echo, are made by impression and reaction; yet I cannot in my simplicity conceive it, how bodies that come not near or touch each other can make a figure by impression and reaction. They say it proceeds from the motions of the medium of light or air, or both, *viz.* that the medium is like a long stick with two ends, whereof one touches the object, the other the organ of sense, and that one end of it moving, the other moves also at the same point of time, by which motions it may make several figures. But I cannot conceive how this motion of pressing forward and backward should make [81] so many figures, wherein there is so much variety and curiosity.

But, say light and air are as one figure, and like as a seal do print another body; I answer, if anything could print, yet it is not probable that so soft and rare bodies as light and air could print such solid bodies as glass, nor could air by reverberation make such a sound as echo. But mistake me not, for I do not say that the corporeal motions of light or air cannot or do not pencil, copy, or pattern out any figure, for both light and air are very active in such sorts of motions, but I say they cannot do it on any other bodies but their own.

But to cut off tedious and unnecessary disputes, I return to the expressing of my own opinion, and believe that the glass in its own substance does figure out the copy of the face or the like, and from that copy the sensitive motions in the eyes take another copy, and so the rational from the sensitive; and in this manner is made both rational and sensitive perception, sight, and knowledge. The same with echoes; for the air patterns out the copy of the sound, and then the sensitive corporeal motions in the ear pattern again this copy from the air, and so do make the perception and sense of hearing.

You may ask me, Madam, if it be so, that the glass and the air copy out the figure of the face and of sound, whether the glass may be said to see and the air to speak. I answer, I cannot tell that; for though I say that the air repeats the words, and glass represents the face, yet I cannot guess what their perceptions are, only this I may say, that the air has an elemental, and the glass a mineral, but not an animal perception.

But if these figures were made by the pressures of several objects or parts and by reaction, there could not be such variety [82] as there is, for they could but act by one sort of motion. Likewise it is improbable that sounds, words, or voices should like a company of wild geese fly into the air, and so enter into the ears of the hearers, as they into their nests. Neither can I conceive how in this manner a word can enter so many ears, that is, be divided into every ear, and yet strike every ear with an undivided vocal sound. You will say, as a small fire does heat and warm all those that stand by; for the heat issues from the fire as the light from the sun. I answer, all what issues and has motion has a body, and yet most learned men deny that sound, light, and heat have bodies. But if they grant of light that it has a body, they say it moves and presses the air, and the air the eye, and so of heat; which if so, then the air must not move to any other motion but light, and only to one sort of light, as the sun's light; for if it did move in any other motion, it would disturb the light. For if a bird did but fly in the air, it would give all the region of air another motion, and so put out or alter the light, or at least disturb it; and wind would make a great disturbance in it.

Besides, if one body did give another body motion, it must needs give it also substance, for motion is either something or nothing, body or no body, substance or no substance; if nothing, it cannot enter into another body; if something, it must lessen the bulk of the body it quits, and increase the bulk of the body it enters, and so the sun and fire, with giving light and heat, would become less, for they cannot both give and keep at once, for this is as impossible as for a man to give to another creature his human nature and yet to keep it still.

Wherefore my opinion is for heat, that [83] when many men stand round about a fire and are heated and warmed by it, the fire does not give them anything, nor do they receive something from the fire, but the sensitive motions in their bodies pattern out the object of the fire's heat, and so they become more or less hot according as their patterns are numerous or perfect. And as for air, it patterns out the light of the sun, and the sensitive motions in the eyes of animals pattern out the light in the air. The like for echoes or any other sound, and for the figures which are presented in a looking-glass. And thus millions of parts or creatures may make patterns of one or more objects, and the objects neither give nor lose anything. And this I repeat here, that my meaning of perception may be the better understood, which is the desire of, Madam,

Your faithful friend and servant.

25.

Madam,

I perceive you are not fully satisfied with my former letter concerning echo and a figure presented in a looking-glass [. . .]

[84] [. . .] And as for the figure presented in a looking-glass, I cannot conceive it to be made by pressure and reaction; for although there is both pressure and reaction in nature, and those very frequent amongst nature's parts, yet they do neither make perception nor production, although both pressure and reaction are made by corporeal self-motions. Wherefore the figure presented in a looking-glass or any other smooth glassy body is, in my opinion, only made by the motions of the looking-glass, which do both pattern out and present the figure of an external object in the glass.

But you will say, why do not the motions of other bodies pattern out and present the figures of external objects as well as smooth glassy bodies do? I answer, they may pattern out external objects, for anything I know; but the reason that their figures are not presented to our eyes lies partly in the presenting subject itself, partly [85] in our sight; for it is observed that two things are chiefly required in a subject that will present the figure of an external object; first it must be smooth, even, and glassy; next it must not be transparent. The first is manifest by experience; for the subject being rough and uneven will never be able to present such a figure; as for example, a piece of steel rough and unpolished, although it may perhaps pattern out the figure of an external object, yet it will never present its figure, but as soon as it is polished and made smooth and glassy, the figure is presently perceived. But this is to be observed, that smooth and glassy bodies do not always pattern out exterior objects exactly, but some better, some worse; like as painters have not all the same ingenuity; neither do all eyes pattern out all objects exactly; which proves that the perception of sight is not made by pressure and reaction, otherwise there would be no difference, but all eyes would see alike.

Next, I say, it is observed that the subject which will present the figure of an external object must not be transparent; the reason is that the figure of light, being a substance of a piercing and penetrating quality, has more force on transparent than on other solid dark bodies, and so disturbs the figure of an external object patterned out in a transparent body, and quite over-masters it. But you will say, you have found by experience that if you hold a burning candle before a transparent glass, although it be in an open sunlight, yet the figure of light and flame of the candle will clearly be seen in the glass. I answer, that it is another thing with the figure of candle-light, than of a duskish or dark body; for a candle-light, though it is not of the same [86] sort as the sun's light, yet it is the same nature and quality, and therefore the candle-light does resist and oppose the light of the sun, so that it cannot have so much power over it as over the figures of other bodies patterned out and presented in transparent glass.

Lastly, I say that the fault often-times lies in the perceptive motions of our sight, which is evident by a plain and concave glass; for in a plain looking-glass, the further you go from it, the more your figure presented in the glass seems to draw backward; and in a concave glass, the nearer you go to it, the more seems your figure to come forth. Which effects are like as a house or tree appears to a traveller; for as the man moves from the house or tree, so the house or tree seems to move from the man; or like one that sails upon a ship, who imagines that the ship stands still and the land moves; when as yet it is the man and the ship that moves, and not the house, or tree, or the land. So when a man turns round in a quick motion, or when his head is dizzy, he imagines the room or place where he is turns round. Wherefore it is the inherent perceptive motions in the eye, and not the motions in the looking-glass, which cause these effects.

And as for several figures that are presented in one glass, it is absurd to imagine that so many several figures made by so many several motions should touch the eye; certainly this would make such a disturbance if all figures were to enter or but to touch the eye, as the eye would not perceive any of them, at least not distinctly. Wherefore it is most probable that the glass patterns out those figures, and the sensitive corporeal motions in the eye take again a pattern from those figures patterned out by the glass, and so make [87] copies of copies; but the reason why several figures are presented in one glass in several places is that two perfect figures cannot be in one point, nor made by one motion, but by several corporeal motions.

Concerning a looking-glass made in the form or shape of a cylinder, why it represents the figure of an external object in another shape and posture than the object is, the cause is the shape and form of the glass, and not the patterning motions in the glass. But this discourse belongs properly to the optics, wherefore I will leave it to those that are versed in that art to enquire and search more after the rational truth thereof. In the meantime, my opinion is that though the object is the occasion of the figure presented in a looking-glass, yet the figure is made by the motions of the glass or body that presents it, and that the figure of the glass perhaps may be patterned out as much by the motions of the object in its own substance, as the figure of the object is patterned out and presented by the motions of the glass in its own body or substance. And thus I conclude and rest, Madam,

Your faithful friend and servant.

28.

Madam,

From sound I am come to scent, in the discourse whereof, your author is pleased to set down these following propositions: "1. That smelling is hindered by cold and helped by heat; 2. That when the wind blows from [the direction of] the object, the smell is the stronger, and when it blows from the sentient towards the object, the weaker, which by experience is found in dogs, that follow the track of beasts by the scent; 3. That such bodies as are least pervious[92] to the fluid medium, yield less smell than such as are more pervious; 4. That such bodies as are of their own nature odorous, become yet more odorous, when they are bruised; 5. That when the breath is stopped (at least in man) nothing can be smelt; 6. That the sense of smelling is also taken away by the stopping of the nostrils, though the mouth be left open."[93]

To begin from the last, I say that the nose is like the other sensitive organs, which if they be stopped, the corporeal sensitive motions cannot take copies of the exterior objects, and therefore [92] must alter their action of patterning to some other, for when the eye is shut and cannot perceive outward objects then it works to the sense of touch, or on the inside of the organ to some phantasms; and so do the rest of the senses.

As for the stopping of breath, why it hinders the scent, the cause is that nostrils and the mouth are the chief organs to receive air and to let out breath. But though they be common passages for air and breath, yet taste is only made in the mouth and tongue, and scent in the nose; not by the pressure of meat and the odiferous object, but the patterning out the several figures or objects of scent and taste, for the nose and the mouth will smell and taste one, nay several things at the same time, like as the eye will see light, color, and other objects at once, which I think can hardly be done by pressures; and the reason is that the

---

92. Pervious: permeable.
93. Hobbes, *Elements*, part 4, chapter 29, section 12.

sensitive motions in the sensitive organs make patterns of several objects at one time, which is the cause that when flowers and such like odiferous bodies are bruised, there are as many figures made as there are parts bruised or divided, and by reason of so many figures the sensitive knowledge is stronger.

But that stones, minerals, and the like, seem not so strong to our smell, the reason is that their parts being close and united, the sensitive motions in the organ cannot so readily perceive and pattern them out as those bodies which are more porous and divided.

But as for the wind blowing the scent either to or from the sentient, it is like a window or door that by the motion of opening and shutting hinders or disturbs the sight; for bodies coming between the object and the organ make a stop of that perception. And as for the dogs smelling out the tracks of beasts, the cause [93] is that the earth or ground has taken a copy of that scent, which copy the sensitive motions in the nose of the dog do pattern out, and so long as that figure or copy lasts, the dog perceives the scent; but if he does not follow or hunt readily, then there is either no perfect copy made by the ground, or otherwise he cannot find it, which causes him to seek and smell about until he has it. And thus smell is not made by the motion of the air, but by the figuring motions in the nose.

Wherefore it is also to be observed that not only the motions in one, but in millions of noses, may pattern out one little object at a time, and therefore it is not that the object of scent fills a room by sending out the scent from its substance, but that so many figures are made of that object of scent by so many several sensitive motions, which pattern the same out; and so the air or ground, or any other creature whose sensitive motions pattern out the object of scent, may perceive the same, although their sensitive organs are not like to those of animal creatures. For if there be but such sensitive motions and perceptions, it is no matter for such organs.

Lastly, it is to be observed that all creatures have not the same strength of smelling, but some smell stronger, some weaker, according to the disposition of their sensitive motions. Also there be other parts in the body which pattern out the object of scent besides the nose, but those are interior parts, and take their patterns from the nose as the organ properly designed for it; neither is their resentment[94] the same, because their motions are not alike, for the stomach may

---

94. Resentment: sensation.

perceive and pattern out a scent with [94] aversion, when the nose may pattern it out with pleasure. And thus much also of scent, I conclude and rest, Madam,

Your faithful friend and servant.

29.

Madam,

Concerning your learned author's discourse of density and rarity, he defines "thick" "to be that which takes up more parts of a space given; and thin, which contains fewer parts of the same magnitude; not that there is more matter in one place than in another equal place, but a greater quantity of some named body; wherefore the multitude and paucity of the parts contained within the same space do constitute density and rarity."[95]

Whereof my opinion is that there is no more nor less space than body according to its dilation or contraction, and that space and place are dilated and contracted with the body, according to the magnitude of the body, for body, place, and magnitude are the same thing, only place is in regard of the several parts of the body, and there is as well space between things distant a hair's breadth from one another as between things distant a million of miles, but yet this space is nothing from the body; but it makes that that body [95] has not the same place with this body, that is, that this body is not that body, and this body's place is not that body's place.

Next your author says he has "already clearly enough demonstrated, that there can be no beginning of motion, but from an external and moved body, and that heavy bodies being once cast upward cannot be cast down again, but by external motion."[96] Truly, Madam, I will not speak of your author's demonstrations, for it is done most by art, which I have no knowledge in, but I think I have probably declared that all the actions of nature are not forced by one part driving, pressing, or shoving another, as a man does a wheelbarrow, or a whip a horse; nor by reactions, as if men were at football or cuffs,[97] or as men with carts meeting each other in a narrow lane.

---

95. Hobbes, *Elements*, part 4, chapter 30, section 1.
96. Hobbes, *Elements*, part 4, chapter 30, section 2.
97. At cuffs: fighting.

But to prove there is no self-motion in nature, he goes on and says, "to attribute to created bodies the power to move themselves, what is it else, than to say that there be creatures which have no dependence upon the Creator?"[98] To which I answer that if man (who is but a single part of nature) has given him by God the power and a free will of moving himself, why should not God give it to nature? Neither can I see how it can take off the dependence upon God, more than eternity; for if there be an eternal creator, there is also an eternal creature, and if an eternal master, an eternal servant, which is nature; and yet nature is subject to God's command and depends upon him; and if all God's attributes be infinite, then his bounty is infinite also, which cannot be exercised but by an infinite gift, but a gift does not cause a less dependence.

I do not say that man has an absolute free will, or power [96] to move according to his desire; for it is not conceived that a part can have an absolute power. Nevertheless his motion both of body and mind is a free and self-motion, and such a self-motion has every thing in nature according to its figure or shape; for motion and figure, being inherent in matter, matter moves figuratively. Yet I do not say that there is no hindrance, obstruction, and opposition in nature; but as there is no particular creature that has an absolute power of self-moving; so that creature which has the advantage of strength, subtlety, or policy,[99] shape, or figure, and the like, may oppose and overpower another which is inferior to it in all this; yet this hindrance and opposition do not take away self-motion. But I perceive your author is much for necessitation and against free will, which I leave to moral philosophers and divines.

And as for the ascending of light and descending of heavy bodies, there may be many causes, but these four are perceivable by our senses, as bulk, or quantity of body; grossness of substance; density; and shape or figure, which make heavy bodies descend. But little quantity, purity of substance, rarity, and figure or shape make light bodies ascend. Wherefore I cannot believe that there are "certain little bodies as atoms, and by reason of their smallness, invisible, differing from one another in consistence,[100] figure, motion, and magnitude, intermingled with the air,"[101] which should be the cause of the descending of heavy bodies. And concerning air, whether it be subject to our senses or not,[102] I say that if air

---

98. Hobbes, *Elements*, part 4, chapter 30, section 2.

99. Policy: political shrewdness, cunning.

100. Consistence: solidity, density.

101. Hobbes, *Elements*, part 4, chapter 30, section 3.

102. Hobbes takes up this question in *Elements*, part 4, chapter 30, section 14, arguing that because air is not detectable by our senses, we know it exists through reasoning.

be neither hot nor cold, it is not subject; but if it be, the sensitive motions will soon pattern it out and declare it.

I'll conclude with your author's question, "what [97] the cause is, that a man does not feel the weight of water in water?"[103] And answer, it is the dilating nature of water. But of this question and of water I shall treat more fully hereafter, and so I rest, Madam,

Your faithful friend and servant.

30.

Madam,

I am now reading the works of that famous and most renowned author, Descartes, out of which I intend to pick out only those discourses which I like best, and not to examine his opinions as they go along from the beginning to the end of his books.[104] And in order to this, I have chosen in the first place his discourse of motion, and do not assent to his opinion when he defines "motion" to be "only a mode of a thing, and not the thing or body itself";[105] for, in my opinion, there can be no abstraction made of motion from body, neither really, nor in the manner of our conception, for how can I conceive that which is not, nor cannot be in nature, that is, to conceive motion without body? Wherefore motion is but one thing with body, without any separation or abstraction soever.

Neither does it agree with my reason that "one body can give or transfer motion into another body, and as much motion it gives [98] or transfers into that body, as much loses it; as for example, in two hard bodies thrown against one another, where one, that is thrown with greater force, takes the other along with it, and loses as much motion as it gives it."[106] For how can motion, being no substance, but only a mode, quit one body and pass into another? One body may either occasion or imitate another's motion, but it can neither give nor take

---

103. Hobbes, *Elements*, part 4, chapter 30, section 6.

104. Cavendish is discussing Descartes's *Principles of Philosophy* (1644).

105. Descartes, *Principles*, part 2, section 25. Descartes's original text says "*non rem aliquam subsistem*," better translated as "not some subsisting thing" rather than "not the thing or body itself." See *Oeuvres de Descartes*, vol. 8, ed. Charles Adam and Paul Tannery (Paris: Vrin, 1996), 54.

106. Descartes, *Principles*, part 2, section 40.

away what belongs to its own or another body's substance, no more than matter can quit its nature from being matter; and therefore my opinion is that if motion does go out of one body into another, then substance goes too; for motion, and substance or body, as aforementioned, are all one thing, and then all bodies that receive motion from other bodies must needs increase in their substance and quantity, and those bodies which impart or transfer motion must decrease as much as they increase.

Truly, Madam, that neither motion nor figure should subsist by themselves and yet be transferrable into other bodies is very strange, and as much as to prove them to be nothing and yet to say they are something. The like may be said of all others which they call accidents, as skill, learning, knowledge, etc., saying, they are no bodies, because they have no extension, but [are] inherent in bodies or substances as in their subjects; for although the body may subsist without them, yet they being always with the body, body and they are all one thing. And so is power and body, for body cannot quit power, nor power the body, being all one thing.

But to return to motion, my opinion is that all matter is partly animate and partly inanimate, and all matter is moving and moved, and that there is no part of nature [99] that has not life and knowledge, for there is no part that has not a commixture of animate and inanimate matter; and though the inanimate matter has no motion, nor life and knowledge of itself, as the animate has, nevertheless being both so closely joined and commixed as in one body, the inanimate moves as well as the animate, although not in the same manner; for the animate moves of itself, and inanimate moves by the help of the animate, and thus the animate is moving and the inanimate moved; not that the animate matter transfers, infuses, or communicates its own motion to the inanimate; for this is impossible, by reason it cannot part with its own nature, nor alter the nature of inanimate matter, but each retains its own nature; for the inanimate matter remains inanimate, that is, without self-motion, and the animate loses nothing of its self-motion, which otherwise it would, if it should impart or transfer its motion into the inanimate matter; but only as I said heretofore, the inanimate works or moves with the animate because of their close union and commixture; for the animate forces or causes the inanimate matter to work with her; and thus one is moving, the other moved, and consequently there is life and knowledge in all parts of nature, by reason in all parts of nature there is a commixture of animate and inanimate matter.

And this life and knowledge is sense and reason, or sensitive and rational corporeal motions, which are all one thing with animate matter without any distinction or abstraction, and can no more quit matter than matter can quit motion. Wherefore every creature being composed of this commixture of animate and inanimate matter has also self-motion, that is life and knowledge, [100] sense and reason, so that no part has need to give or receive motion to or from another part; although it may be an occasion of such a manner of motion to another part, and cause it to move thus or thus. As for example, a watchmaker does not give the watch its motion, but he is only the occasion that the watch moves after that manner, for the motion of the watch is the watch's own motion, inherent in those parts ever since that matter was, and if the watch ceases to move after such a manner or way, that manner or way of motion is nevertheless in those parts of matter the watch is made of, and if several other figures should be made of that matter, the power of moving in the said manner or mode would yet still remain in all those parts of matter as long as they are body and have motion in them.

Wherefore one body may occasion another body to move so or so, but not give it any motion, but every body (though occasioned by another to move in such a way) moves by its own natural motion; for self-motion is the very nature of animate matter, and is as much in hard as in fluid bodies, although your author denies it, saying, "the nature of fluid bodies consists in the motion of those little insensible parts into which they are divided, and the nature of hard bodies, when those little particles joined closely together, do rest";[107] for there is no rest in nature. Wherefore if there were a world of gold and a world of air, I do verily believe that the world of gold would be as much interiously active as the world of air exteriously; for nature's motions are not all external or perceptible by our sense, neither are they all circular, or only of one sort, but there is an infinite change and variety of motions; for though [101] I say in my *Philosophical Opinions*, "as there is but one only matter, so there is but one only motion";[108] yet I do not mean there is but one particular sort of motions, as either circular or straight or the like, but that the nature of motion is one and the same, simple and entire in itself, that is, it is mere motion, or nothing else but corporeal motion; and that as there are infinite divisions or parts of matter, so there are infinite changes and varieties of motions, which is the reason that I call motion as well infinite as matter; first that matter and motion are but one thing, and if matter be infinite, motion must be so too; and secondly, that motion is infinite

---

107. Descartes, *Principles*, part 2, section 54. Strictly speaking, Descartes's claim is that the particles in hard bodies are at rest relative to each other.

108. Cavendish paraphrases *Philosophical and Physical Opinions* (1663), part 1, chapter 5.

in its changes and variations, as matter is in its parts. And thus much of motion for this time; I add no more, but rest, Madam,

Your faithful friend and servant.

31.

Madam,

I observe your author in his discourse of place makes a difference between an interior and exterior place, and that according to this distinction, "one body may be said to change, and not to change its place at the same time, and that one body may succeed into another's place."[109] But I am not of this opinion, for I believe [102] not that there is any more place than body; as for example, water being mixed with earth, the water does not take the earth's place, but as their parts intermix, so do their places, and as their parts change, so do their places, so that there is no more place than there is water and earth; the same may be said of air and water, or air and earth, or did they all mix together; for as their bodies join, so do their places, and as they are separated from each other, so are their places.

Say a man travels a hundred miles, and so a hundred thousand paces; but yet this man has not been in a hundred thousand places, for he never had any other place but his own; he has joined and separated himself from a hundred thousand, nay, millions of parts, but he has left no places behind him. You will say, if he travel the same way back again, then he is said to travel through the same places. I answer, it may be the vulgar[110] way of expression, or the common phrase; but to speak properly, after a philosophical way, and according to the truth in nature, he cannot be said to go back again through the same places he went, because he left none behind him, or else all his way would be nothing but place after place, all the hundred miles along. Besides, if "place" should be taken so as to express the joining to the nearest bodies which compass him about, certainly he would never find his places again; for the air, being fluid, changes or moves continually, and perchance the same parts of the air which compassed him once will never come near him again.

---

109. Descartes, *Principles*, part 2, sections 10, 11, 13, and 14.
110. Vulgar: common, ordinary.

But you may say, if a man be hurt or has some mischance in his body, so as to have a piece of flesh cut out and new flesh growing there; then we say, because the adjoining parts do [103] not change, that a new piece of flesh is grown in the same place where the former flesh was, and that the place of the former flesh cut or fallen out is the same of this new grown flesh. I answer, in my opinion, it is not, for the parts not being the same, the places are not, but every one has its own place. But if the wound be not filled or closed up with other new flesh, you will say that according to my opinion there is no place then at all. I say, yes, for the air or anything else may be there as new parts joining to the other parts; nevertheless, the air, or that same body which is there, has not taken the flesh's place which was there before, but has its own; but, by reason the adjoining parts remain, man thinks the place remains there also, which is no consequence. 'Tis true, a man may return to the same adjoining bodies where he was before, but then he brings his place with him again, and as his body, so his place returns also; and if a man's arm be cut off, you may say there was an arm heretofore, but you cannot say properly, this is the place where the arm was.

But to return to my first example of the mixture of water and earth or air; suppose water is not porous, but only dividable, and has no other place but what is its own body's, and that other parts of water intermix with it by dividing and composing; I say, there is no more place required than what belongs to their own parts, for if some contract, others dilate, some divide, others join, the places are the same according to the magnitude of each part or body.

The same may be said of all kinds or sorts of mixtures, for one body has but one place; and so if many parts of the same nature join into one body and increase the bulk of the body, [104] the place of that same body is [increased] accordingly; and if they be bodies of different natures which intermix and join, each several keeps its place. And so each body and each particular part of a body has its place, for you cannot name body or part of a body, but you must also understand place to be with them, and if a point should dilate to a world, or a world contract to a point, the place would always be the same with the body. And thus I have declared my opinion on this subject, which I submit to the correction of your better judgment, and rest, Madam,

Your Ladyship's faithful friend and humble servant.

32.

Madam,

In my last, I hope I have sufficiently declared my opinion that to one body belongs but one place, and that no body can leave a place behind it, but wheresoever is body, there is place also. Now give me leave to examine this question: when a body's figure is printed on snow or any other fluid or soft matter, as air, water, and the like, whether it be the body that prints its own figure upon the snow, or whether it be the snow that patterns the figure of the body?

My answer is that it is not the body which [105] prints its figure upon the snow, but the snow that patterns out the figure of the body; for if a seal be printed upon wax, 'tis true, it is the figure of the seal which is printed on the wax, but yet the seal does not give the wax the print of its own figure, but it is the wax that takes the print or pattern from the seal, and patterns or copies it out in its own substance, just as the sensitive motions in the eye do pattern out the figure of an object, as I have declared heretofore.

But you will say, perhaps, a body being printed upon snow, as it leaves its print, so it leaves also its place with the print in the snow. I answer, that does not follow; for the place remains still the body's place, and when the body removes out of the snow, it takes its place along with it. Just like a man whose picture is drawn by a painter, when he goes away, he leaves not his place with his picture, but his place goes with his body; and as the place of the picture is the place of the color or paint, and the place of the copy of an exterior object patterned out by the sensitive corporeal motions is the place of the sensitive organ, so the place of the print in snow is the snow's place; or else, if the print were the body's place that is printed, and not the snow's, it might as well be said that the motion and shape of a watch were not the motion and shape of the watch, but of the hand of him that made it.

And as it is with snow, so it is with air, for a man's figure is patterned out by the parts and motions of the air wheresoever he moves; the difference is only that air being a fluid body does not retain the print so long as snow or a harder body does, but when the body removes, the print is presently dissolved. But I wonder much, your author denies [106] that there can be two bodies in one place, and yet makes two places for one body, when all is but the motions of one body. Wherefore a man sailing in a ship cannot be said to keep place and to change his place; for it is not place he changes, but only the adjoining parts, as leaving some and joining to others; and it is very improper to attribute that to place which belongs to parts, and to make a change of place out of change of parts.

I conclude, repeating once again that figure and place are still remaining the same with body. For example, let a stone be beat to dust, and this dust be severally dispersed, nay, changed into numerous figures; I say, as long as the substance of the stone remains in the power of those dispersed and changed parts, and their corporeal motions, the place of it continues also; and as the corporeal motions change and vary, so do place, magnitude, and figure, together with their parts or bodies, for they are but one thing. And so I conclude, and rest, Madam,

Your faithful friend and servant.

[107]   **33.**

Madam,

I am absolutely of your author's opinion when he says "that all bodies of the universe are of one and the same matter, really divided into many parts, and that these parts are diversely moved."[111] But that these motions should be circular more than of any other sort, I cannot believe, although he thinks that this is the most probable way to find out the causes of natural effects. For nature is not bound to one sort of motions more than to another, and it is but in vain to endeavor to know how and by what motions God did make the world, since creation is an action of God, and God's actions are incomprehensible.

Wherefore his aethereal whirlpools and little particles of matter, which he reduces to three sorts and calls them the three elements of the universe, their circular motions, several figures, shavings, and many the like, which you may better read than I rehearse to you, are, to my thinking, rather fancies than rational or probable conceptions.[112] For how can we imagine that the universe was set a-moving as a top by a whip, or a wheel by the hand of a spinster, and that the vacuities were filled up with shavings? For these violent motions would rather have disturbed and disordered nature; and though nature uses variety in her

---

111. Descartes, *Principles*, part 3, section 46.

112. Descartes maintains that the visible universe is composed of three kinds of particles, which he calls "elements" (*Principles*, part 3, section 52), all in circular motion (*Principles*, part 2, section 39, and part 3, section 46).

motions or actions, yet these are not extravagant, nor by force or violence, but orderly, temperate, free, and easy, which causes me [108] to believe the earth turns about rather than the sun; and though corporeal motions for variety make whirlwinds, yet whirlwinds are not constant.

Neither can I believe that the swiftness of motion could make the matter more subtle and pure than it was by nature, for it is the purity and subtlety of the matter that causes motion and makes it swifter or slower, and not motion the subtlety and purity of matter; motion being only the action of matter; and the self-moving part of matter is the working part of nature, which is wise, and knows how to move and form every creature without instruction; and this self-motion is as much her own as the other parts of her body, matter, and figure, and is one and the same with herself, as a corporeal, living, knowing, and inseparable being, and a part of herself.

As for the several parts of matter, I do believe that they are not all of one and the same bigness, nor of one and the same figure, neither do I hold their figures to be unalterable; for if all parts of nature be corporeal, they are dividable, composable, and intermixable, and then they cannot be always of one and the same sort of figure; besides, nature would not have so much work if there were no change of figures. And since her only action is change of motion, change of motion must needs make change of figures. And thus natural parts of matter may change from lines to points, and from points to lines, from squares to circles, and so forth, infinite ways, according to the change of motions; but though they change their figures, yet they cannot change their matter; for matter as it has been, so it remains constantly in each degree, as the rational, sensitive, and inanimate, none becomes purer, none grosser [109] than ever it was, notwithstanding the infinite changes of motions which their figures undergo; for motion changes only the figure, not the matter itself, which continues still the same in its nature, and cannot be altered without a confusion or destruction of nature. And this is the constant opinion of, Madam,

Your faithful friend and humble servant.

[111]   **35.**

Madam,

"That the mind," according to your author's opinion, "is a substance really distinct from the body, and may be actually separated from it and subsist without it,"[113] if he means the natural mind and soul of man, not the supernatural or divine, I am far from his opinion; for though the mind moves only in its own parts, and not upon or with the parts of inanimate matter, yet it cannot be separated from these parts of matter and subsist by itself, as being a part of one and the same matter the inanimate is of (for there is but one only matter, and one kind of matter, although of several degrees), only it is the self-moving part; but yet this cannot empower it to quit the same natural body whose part it is.

Neither can I apprehend that the mind's or soul's seat should be in the "glandula" or kernel of the brain,[114] and there sit like a spider in a cobweb, to whom the least motion of the cobweb gives intelligence of a fly, which he is ready to assault, and that the brain should get intelligence by the animal spirits[115] as his servants, which run to and fro like ants to inform it; or that the mind should, according to others' opinions, be a light, and embroidered all with ideas, like a herald's coat; and that the sensitive organs should have no knowledge in themselves, but serve only like peeping-holes for the mind, or barn-doors to receive bundles of pressures, like sheaves of corn. For there being a [112] thorough mixture of animate, rational and sensitive, and inanimate matter, we cannot assign a certain seat or place to the rational, another to the sensitive, and another to the inanimate, but they are diffused and intermixed throughout all the body. And this is the reason that sense and knowledge cannot be bound only to the head or brain.

But although they are mixed together, nevertheless they do not lose their interior natures by this mixture, nor their purity and subtlety, nor their proper motions or actions, but each moves according to its nature and substance, without confusion. The actions of the rational part in man, which is the mind or soul, are called "thoughts" or thoughtful perceptions, which are numerous, and

---

113. Cavendish doesn't cite her source, but she is probably paraphrasing part 4 of Descartes's *Discourse on Method*, for this is the text she cites in the next letter, and there is no evidence that she had the *Meditations* translated for her.

114. This is a reference to the pineal gland, which Descartes thought was the part of the brain to and from which "animal spirits" moved throughout the body, and to which the immaterial soul was joined.

115. Animal spirits were thought to be a thin, rarefied physical substance that flowed through tubular nerves, as blood flows through veins and arteries.

so are the sensitive perceptions; for though man or any other animal has but five exterior sensitive organs, yet there be numerous perceptions made in these sensitive organs, and in all the body; nay, every several pore of the flesh is a sensitive organ, as well as the eye or the ear. But both sorts, as well the rational as the sensitive, are different from each other, although both do resemble another, as being both parts of animate matter, as I have mentioned before. Wherefore I'll add no more, only let you know that I constantly remain, Madam,

Your faithful friend and servant.

[113]   **36.**

Madam,

That all other animals besides man want reason, your author endeavors to prove in his *Discourse on Method*,[116] where his chief argument is that other animals cannot express their mind, thoughts, or conceptions, either by speech or any other signs as man can do. For, says he, "it is not for want of the organs belonging to the framing of words, as we may observe in parrots and pies,[117] which are apt enough to express words they are taught, but understand nothing of them."[118]

My answer is that one man expressing his mind by speech or words to another does not declare by it his excellency and supremacy above all other creatures, but for the most part more folly, for a talking man is not so wise as a contemplating man. But by reason other creatures cannot speak or discourse with each other as men, or make certain signs whereby to express themselves as dumb and deaf men do, should we conclude they have neither knowledge, sense, reason, or intelligence? Certainly, this is a very weak argument; for one part of a man's body, as one hand, is not less sensible than the other, nor the heel less sensible than the heart, nor the leg less sensible than the head, but each part has its sense and reason, and so consequently its sensitive and rational knowledge; and although they cannot talk or give intelligence to each other by speech, nevertheless each has its own peculiar and [114] particular knowledge, just as each particular man has his own particular knowledge, for one man's knowledge is not another man's

---

116. On Cavendish's source for her quotations, see the Introduction (p. xxxi).

117. Pies: magpies.

118. See Descartes, *Discourse on Method*, part 5.

knowledge; and if there be such a peculiar and particular knowledge in every several part of one animal creature as man, well may there be such in creatures of different kinds and sorts. But this particular knowledge belonging to each creature does not prove that there is no intelligence at all between them, no more than the want of human knowledge does prove the want of reason; for reason is the rational part of matter, and makes perception, observation, and intelligence different in every creature and every sort of creatures, according to their proper natures, but perception, observation, and intelligence do not make reason, reason being the cause and they the effects.

Wherefore though other creatures have not the speech, nor mathematical rules and demonstrations, with other arts and sciences, as men [do], yet may their perceptions and observations be as wise as men's, and they may have as much intelligence and commerce between each other, after their own manner and way, as men have after theirs. To which I leave them, and man to his conceited prerogative and excellence, resting, Madam,

Your faithful friend and servant.

[115]   **37.**

Madam,

Concerning sense and perception, your author's opinion is that it is made by "a motion or impression from the object upon the sensitive organ, which impression by means of the nerves, is brought to the brain, and so to the mind or soul, which only perceives in the brain."[119] Explaining it by the example of a man being blind or walking in dark, who by the help of his stick can perceive when he touches a stone, a tree, water, sand, and the like; which example he brings to make a comparison with the perception of light. "For," says he, "light in a shining body is nothing else but a quick and lively motion or action, which through the air and other transparent bodies tends towards the eye, in the same manner as the motion or resistance of the bodies the blind man meets withal, tend through the stick towards the hand; wherefore it is no wonder that the sun can display its rays so far in an instant, seeing that the same action, whereby one end of the

---

119. This is a very rough paraphrase of Descartes, *Principles*, part 4, section 189.

stick is moved, goes instantly also to the other end, and would do the same if the stick were as long as heaven is as distant from Earth."[120]

To which I answer, first, that it is not only the mind that perceives in the kernel of the brain, but that there is a double perception, rational and sensitive, and that the mind perceives by the rational, but the body and the sensitive organs by the sensitive perception; and as there is a double perception, so there is also a double knowledge, [116] rational and sensitive, one belonging to the mind, the other to the body; for I believe that the eye, ear, nose, tongue, and all the body have knowledge as well as the mind; only the rational matter, being subtle and pure, is not encumbered with the grosser part of matter to work upon or with it, but leaves that to the sensitive, and works or moves only in its own substance, which makes a difference between thoughts and exterior senses.

Next I say that it is not the motion or reaction of the bodies the blind man meets withal which makes the sensitive perception of these objects, but the sensitive corporeal motions in the hand do pattern out the figure of the stick, stone, tree, sand, and the like. And as for comparing the perception of the hand, when by the help of the stick it perceives the objects, with the perception of light, I confess that the sensitive perceptions do all resemble each other, because all sensitive parts of matter are of one degree, as being sensible parts, only there is a difference according to the figures of the objects presented to the senses; and there is no better proof for perception being made by the sensitive motions in the body, or sensitive organs, but that all these sensitive perceptions are alike and resemble one another; for if they were not made in the body of the sentient, but by the impression of exterior objects, there would be so much difference between them, by reason of the diversity of objects, as they would have no resemblance at all.

But for a further proof of my own opinion, did the perception proceed merely from the motion, impression, and resistance of the objects, the hand could not perceive those objects unless they touched the hand itself, as the stick does; for it is not probable [117] that the motions of the stone, water, sand, etc. should leave their bodies and enter into the stick, and so into the hand; for motion must be either something or nothing; if something, the stick and the hand would grow bigger, and the objects touched [would become] less, or else the touching and the touched must exchange their motions, which cannot be done so suddenly,

---

120. Descartes, *Optics*, First Discourse.

especially between solid bodies. But if motion has no body, it is nothing, and how nothing can pass or enter or move some body, I cannot conceive. 'Tis true there is no part that can subsist singly by itself without dependence upon each other, and so parts do always join and touch each other, which I am not against; but only I say perception is not made by the exterior motions of exterior parts of objects, but by the interior motions of the parts of the body sentient. But I have discoursed hereof before, and so I take my leave, resting, Madam,

Your faithful friend and servant.

[120]   **39.**

Madam,

Concerning vapor, clouds, wind, and rain, I am of your author's opinion that "water is changed into vapor, and vapor into air, and that dilated vapors make wind, and condensed vapors, clouds and mists."[121] But I am not for his little particles, "whereof," he says, "vapors are made, by the motion of a rare and subtle matter in the pores of terrestrial bodies";[122] which certainly I should conceive to be loose atoms, did he not make them of several figures and magnitude. For in my opinion there are no such things in nature, which like little flies or bees do fly up into the air; and although I grant that in nature are several parts, whereof some are more rare, others more dense, according to the several degrees of matter, yet they are not single, but all mixed together in one body, and the change of motions in those joined parts is the cause of all changes of figures whatever, without the assistance of any foreign parts.

And thus water of itself is changed to snow, ice, or hail, by its inherent figurative motions; that is, the circular dilation of water by contraction changes into the figure of snow, ice, or hail; or by rarifying motions it turns into the figure of vapor, and this vapor again by contracting motions into the figure of

121. Descartes, *Meteorology*. Cavendish is vague about where she finds these claims, citing "chapters 2, 4, 5, 6." In Descartes's Second Discourse, "Of Vapors and Exhalations," he says that vapors are made of the same particles as liquid water; in the Fourth Discourse, "Of Winds," he says that as the particles become more rarefied, they turn to air, and that further expansions cause winds; and in the Fifth, "Of Clouds," he characterizes clouds and mist as the result of water vapors that have become denser.

122. See Descartes, *Meteorology*, Second Discourse.

hoar-frost;[123] and when all these motions change again into the former, then the figure of ice, snow, hail, vapor, and frost turns again into the figure of [121] water. And this in all sense and reason is the most facile and probable way of making ice, snow, hail, etc. As for rarefaction and condensation, I will not say that they may be forced by foreign parts, but yet they are made by change and alteration of the inherent motions of their own parts, for though the motions of foreign parts may be the occasion of them, yet they are not the immediate cause or actors thereof.

And as for thunder, that clouds of ice and snow, the uppermost being condensed by heat and so made heavy, should fall upon another and produce the noise of thunder[124] is very improbable; for the breaking of a little small string will make a greater noise than a huge shower of snow with falling, and as for ice being hard, it may make a great noise, one part falling upon another, but then their weight would be as much as their noise, so that the clouds or roves[125] of ice would be as soon upon our heads, if not sooner, as the noise in our ears; like as a bullet shot out of a cannon, we may feel the bullet as soon as we hear the noise.

But to conclude, all densations[126] are not made by heat, nor all noises by pressures, for sound is oftener made by division than pressure, and densation by cold than by heat. And this is all for the present, from, Madam,

Your faithful friend and servant.

[122]   **40.**

Madam,

I cannot perceive the rational truth of your author's opinion concerning colors "made by the agitation of little spherical bodies of an aethereal matter, transmitting the action of light";[127] for if colors were made after this manner, there

---

123. Hoar-frost: frozen dew.

124. See Descartes, *Meteorology*, Seventh Discourse, where he explains thunder in this way.

125. Cavendish's use of "rove" appears to be idiosyncratic. She may mean "roofs," since she is describing ice in the atmosphere.

126. Densations: thickenings.

127. Cavendish does not cite her source, but she is probably referring to a claim Descartes makes in *Meteorology*, Eighth Discourse.

would, in my opinion, not be any fixed or lasting color, but one color would be so various, and change faster than every minute; the truth is, there would be no certain or perfect color at all.

Wherefore it seems altogether improbable that such liquid, rare, and disunited bodies should either keep or make inherent and fixed colors; for liquid and rare bodies, whose several parts are united into one considerable bulk of body, their colors are more apt to change than the colors of those bodies that are dry, solid, and dense; the reason is that rare and liquid bodies are more loose, slack, and agile than solid and dry bodies, in so much as in every alteration of motion their colors are apt to change. And if united rare and liquid bodies be so apt to alter and change, how is it probable that those bodies which are small and not united should either keep or make inherent fixed colors? I will not say but that such little bodies may range into such lines and figures as make colors, but then they cannot last, being not united into a lasting body, that is, into a solid, substantial body, proper to make such figures as colors.

But I desire [123] you not to mistake me, Madam, for I do not mean that the substance of colors is a gross thick substance, for the substance may be as thin and rare as flame or light, or in the next degree to it; for certainly the substance of light and the substance of colors come in their degrees very near each other; but according to the contraction of the figures, colors are paler or deeper, or more or less lasting. And as for the reason why colors will change and rechange, it is according as the figures alter or recover their forms; for colors will be as animal creatures, which sometimes are faint, pale, and sick, and yet recover; but when as a particular color is, as I may say, quite dead, then there is no recovering of it.

But colors may seem altered sometimes in our eyes, and yet not be altered in themselves; for our eyes, if perfect, see things as they are presented; and for proof, if any animal should be presented in an unusual posture or shape, we could not judge of it; also if a picture which must be viewed sidewards should be looked upon forward, we could not know what to make of it; so the figures of colors, if they be not placed rightly to the sight, but turned topsy-turvy as the phrase is, or upside-down, or be moved too quick, and this quick motion do make a confusion with the lines of light, we cannot possibly see the color perfectly.

Also, several lights or shades may make colors appear otherwise than in themselves they are, for some sorts of lights and shades may fall upon the substantial figures of colors in solid bodies, in such lines and figures as they may overpower the natural or artificial inherent colors in solid bodies, and for a time make other colors; and many times the lines of light or [124] of shadows will meet and sympathize so with inherent colors, and place their lines so exactly, as they will make those inherent colors more splendorous than in their own nature they are, so that light and shadows will add or diminish or alter colors very much.

Likewise some sorts of colors will be altered to our sight, not by all, but only by some sorts of light, as for example, blue will seem green, and green blue by candle-light, when as other colors will never appear changed, but show constantly as they are; the reason is because the lines of candle-light fall in such figures upon the inherent colors, and so make them appear according to their own figures. Wherefore it is only the alteration of the exterior figures of light and shadows that make colors appear otherwise, and not a change of their own natures.

And hence we may rationally conclude that several lights and shadows by their spreading and dilating lines may alter the face or outside of colors, but not suddenly change them, unless the power of heat, and continuance of time, or any other cause, do help and assist them in that work of metamorphosing or trans- forming of colors; but if the lines of light be only, as the phrase is, skin-deep, that is, but lightly spreading and not deeply penetrating, they may soon wear out or be rubbed off; for though they hurt, yet they do not kill the natural color, but the color may recover and reassume its former vigor and luster. But time and other accidental causes will not only alter but destroy particular colors as well as other creatures, although not all after the same manner, for some will last longer than others. And thus, Madam, there are three sorts of color, natural, artificial, and accidental; but I have discoursed of this subject more at large in my *Philosophical Opinions*,[128] to which I refer you, and rest, Madam,

Your faithful friend and servant.

---

[127] **42.**

Madam,

To conclude my discourse upon the opinions of these two famous and learned authors, which I have hitherto sent you in several letters, I could not choose but repeat the ground of my own opinions in this present; which I desire you to observe well, lest you mistake anything whereof I have formerly discoursed.

First, I am for self-moving matter, which I call the sensitive and rational mat- ter, and the perceptive and architectonical part of nature, which is the life and

---

128. Cavendish discusses colors in *Philosophical and Physical Opinions* (1663), part 5, chapter 41.

knowledge of nature. Next I am of an opinion that all perception is made by corporeal, figuring self-motions, and that the perception of foreign objects is made by patterning them out; as for example, the sensitive perception of foreign objects is by making or taking copies from these objects, so as the sensitive corporeal motions in the eyes copy out the objects of sight, and the sensitive corporeal motions in the ears copy out the objects of sound; the sensitive corporeal motions in the nostrils [128] copy out the objects of scent; the sensitive corporeal motions in the tongue and mouth copy out the objects of taste, and the sensitive corporeal motions in the flesh and skin of the body copy out the foreign objects of touch; for when you stand by the fire, it is not that the fire or the heat of the fire enters your flesh, but that the sensitive motions copy out the objects of fire and light.

As for my book of philosophy, I must tell you, that it treats more of the production and architecture of creatures than of their perceptions, and more of the causes than the effects, more in a general than peculiar way, which I thought necessary to inform you of, and so I remain, Madam,

Your faithful friend and servant.

[132]   45.

Madam,

I understand by your last that you are very desirous to know whether there be not in nature such animal creatures, both for purity and size, as we are not capable to perceive by our sight. Truly, Madam, in my opinion it is very probable there may be animal creatures of such rare bodies as are not subject to our exterior senses, as well as there are elements which are not subject to all our exterior senses; as for example, fire is only subject to our sight and feeling, and not to any other [133] sense, water is subject to our sight, taste, touch, and hearing, but not to smelling; and earth is subject to our sight, taste, touch, and smelling, but not to our hearing; and vapor is only subject to our sight, and wind only to our hearing; but pure air is not subject to any of our senses, but only known by its effects. And so there may likewise be animal creatures which are not subject to any of our senses, both for their purity and life; as for example, I have seen pumped out of a water pump small worms which could hardly be discerned but by a bright sunlight, for they were smaller than the smallest hair, some of a pure scarlet color and some white; but though they were the smallest creatures that ever I did see, yet they were more agile and fuller of life than many a creature of a bigger size, and so small they were, as I am confident they were neither subject to taste, smell, touch, nor hearing, but only to sight, and that neither without difficulty, requiring both a sharp sight and a clear light to perceive them; and I do verily believe that these small animal creatures may be great in comparison to others which may be in nature.

But if it be probable that there may be such small animal creatures in nature as are not subject to our exterior senses by reason of their littleness, it is also probable that there may be such great and big animal creatures in nature as are beyond the reach and knowledge of our exterior senses; for bigness and small-ness are not to be judged by our exterior senses only; but as sense and reason inform us that there are different degrees in purity and rarity, so also in shapes, figures, and sizes in all natural creatures.

Next you desired to know "whether there can be an artificial life, or a life made by art?" [134] My answer is, not; for although there is life in all nature's parts, yet not all the parts are life, for there is one part of natural matter which in its nature is inanimate or without life, and though natural life does produce art, yet art cannot produce natural life, for though art is the action of life, yet it is not life itself. Not but that there is life in art, but not art in life, for life is natural and not artificial; and thus the several parts of a watch may have sense and reason according to the nature of their natural figure, which is steel, but not as they have an artificial shape, for art cannot put life into the watch, life being only natural, not artificial.

Lastly your desire was to know "whether a part of matter may be so small, as it cannot be made less?" I answer, there is no such thing in nature as biggest or least, nature being infinite as well in her actions as in her substance; and I have mentioned in my book of philosophy,[129] and in a letter I sent you heretofore[130] concerning infinite, that there are several sorts of infinites, as infinite in quantity or bulk, infinite in number, infinite in quality, as infinite degrees of hardness, softness, thickness, thinness, swiftness, slowness, etc., as also infinite composi-tions, divisions, creations, dissolutions, etc. in nature; and my meaning is that all these infinite actions do belong to the infinite body of nature, which being infinite in substance must also of necessity be infinite in its actions; but although these infinite actions are inherent in the power of the infinite substance of

---

129. See *Philosophical and Physical Opinions* (1663), part 1, chapter 8.

130. See in this volume, section 1, letter 2.

nature, yet they are never put in act in her parts, by reason there being contraries in nature, and every one of the aforementioned actions having its opposite, they do hinder and obstruct each other so, that none can [135] actually run into infinite; for the infinite degrees of compositions hinder the infinite degrees of divisions; and the infinite degrees of rarity, softness, swiftness, etc. hinder the infinite degrees of density, hardness, slowness, etc., all which nature has ordered with great wisdom and prudence to make an amiable combination between her parts; for if but one of these actions should run into infinite, it·would cause a horrid confusion between nature's parts, nay an utter destruction of the whole body of nature, if I may call it whole. As for example, if one part should have infinite compositions, without the hindrance or obstruction of division, it would at last mount and become equal to the infinite body of nature, and so from a part change to a whole, from being finite to infinite, which is impossible.

Wherefore, though nature has an infinite natural power, yet she does not put this power in act in her particulars; and although she has an infinite force or strength, yet she does not use this force or strength in her parts. Moreover, when I speak of infinite divisions and compositions, creations and dissolutions, etc. in nature, I do not mean so much the infinite degrees of compositions and divisions, as the actions themselves to be infinite in number; for there being infinite parts in nature, and every one having its compositions and divisions, creations and dissolutions, these actions must of necessity be infinite too, to wit, in number, according to the infinite number of parts; for as there is an infinite number of parts in nature, so there is also an infinite number and variety of motions, which are natural actions. However, let there be also infinite degrees of these natural actions, in the body or substance of infinite [136] nature; yet, as I said, they are never put in act, by reason every action has its contrary or opposite, which does hinder and obstruct it from running actually into infinite.

And thus I hope you conceive clearly now what my opinion is, that I do not contradict myself in my works, as some have falsely accused me; for they, by misapprehending my meaning, judge not according to the truth of my sense, but according to their own false interpretation, which shows not only a weakness in their understandings and passions, but a great injustice and injury to me, which I desire you to vindicate whenever you chance to hear such accusations and blemishes laid upon my works, by which you will infinitely oblige, Madam,

Your humble and faithful servant.

SECTION 2

1.

Madam,

Being come now to the perusal of the works of that learned author Dr. More, I find that the only design of his book called *Antidote*[1] is to prove the existence of a God, and to refute, or rather convert atheists; which I wonder very much at, considering he says himself that "there is no man under the cope[2] of heaven but believes a God";[3] which if so, what needs there to make so many arguments to no purpose, unless it be to show learning and wit? In my opinion, it were better to convert pagans to be Christians, or to reform irregular Christians to a more pious life, than to prove that which all men believe, which is the way to bring it into question. For certainly, according to the natural light of reason, there is a God, and no man, I believe, does doubt it; for though there [138] may be many vain words, yet I think there is no such atheistical belief amongst mankind, nay, not only amongst men, but also amongst all other creatures; for if nature believes a God, all her parts, especially the sensitive and rational, which are the living and knowing parts, and are in all natural creatures, do the like; and therefore all parts and creatures in nature do adore and worship God, for anything man can know to the contrary; for no question, but nature's soul adores and worships God as well as man's soul; and why may not God be worshipped by all sorts and kinds of creatures as well as by one kind or sort? I will not say the same way, but I believe there is a general worship and adoration of God; for as God is an infinite deity, so certainly he has an infinite worship and adoration, and there is not any part

---

1. Henry More, *An Antidote against Atheism*, 3rd edition (1662).

2. Cope: canopy.

3. More, *Antidote*, book 1, chapter 10, section 5. Cavendish's quotations are not always accurate, but only significant alterations are noted here. In this case, More's text actually says there is "no nation under the cope of Heaven that does not do divine worship to something or other, and in it to God, as they conceive; wherefore according to the ordinary natural light that is in all men, there is a God."

of nature but adores and worships the only omnipotent God, to whom belongs praise and glory from and to all eternity.[4]

For it is very improbable that God should be worshipped only in part and not in whole, and that all creatures were made to obey man and not to worship God, only for man's sake and not for God's worship, for man's spoil and not God's blessing. But this presumption, pride, vain-glory, and ambition of man, proceeds from the irregularity of nature, who, being a servant, is apt to commit errors; and cannot be so absolute and exact in her devotion, adoration, and worship as she ought, nor so well observant of God as God is observing her. Nevertheless, there is not any of her parts or creatures that God is not acknowledged by, though not so perfectly as he ought, which is caused by the irregularities of nature, as I said before. [139] And so God of his mercy have mercy upon all creatures; to whose protection I commend your Ladyship, and rest, Madam,

Your faithful friend and servant.

2.

Madam,

Since I spoke in my last of the adoration and worship of God, you would fain know whether we can have an idea of God? I answer that naturally we may, and really have a knowledge of the existence of God, as I proved in my former letter, to wit, that there is a God, and that he is the author of all things, who rules and governs all things, and is also the God of nature. But I dare not think that naturally we can have an idea of the essence of God, so as to know what God is in his very nature and essence; for how can there be a finite idea of an infinite God?

You may say, as well as of infinite space. I answer, space is relative, or has respect to body, but there is not anything that can be compared to God; for the idea of infinite nature is material, as being a material creature of infinite material nature. You will say, how can a finite part have an idea of infinite nature? I answer, very well, by reason the idea is part of infinite nature, and [140] so of the same kind, as material; but God being an eternal, infinite, immaterial, individable being, no natural creature can have an idea of him.

---

4. That is, for all of eternity.

You will say that the idea of God in the mind is immaterial; I answer, I cannot conceive that there can be any immaterial idea in nature; but be it granted, yet that immaterial is not a part of God, for God is individable and has no parts; wherefore the mind cannot have an idea of God as it has of infinite nature, being a part of nature; for the idea of God cannot be of the essence of God, as the idea of nature is a corporeal part of nature. And though nature may be known in some parts, yet God being incomprehensible, his essence can by no ways or means be naturally known; and this is constantly believed, by, Madam,

Your faithful friend and servant.

3.

Madam,

Although I mentioned in my last that it is impossible to have an idea of God, yet your author is pleased to say that he "will not stick to affirm, that the idea or notion of God is as easy, as any notion else whatsoever, and that we may know as much of him as of anything else in the world."[5] To which I answer that [141] in my opinion God is not so easily to be known by any creature, as man may know himself; nor his attributes so well, as man can know his own natural proprieties. For God's infinite attributes are not conceivable, and cannot be comprehended by a finite knowledge and understanding, as a finite part of nature; for though nature's parts may be infinite in number, and as they have a relation to the infinite whole, if I may call it so, which is infinite nature, yet no part is infinite in itself, and therefore it cannot know so much as whole nature; and God being an infinite deity, there is required an infinite capacity to conceive him; nay, nature herself, although infinite, yet cannot possibly have an exact notion of God, by reason of the disparity between God and herself; and therefore it is not probable, if the infinite servant of God is not able to conceive him, that a finite part or creature of nature, of what kind or sort soever, whether "spiritual," as your author is pleased to name it, or corporeal, should comprehend God.

---

5. Henry More, *Of the Immortality of the Soul* (1662), book 1, chapter 4, section 1.

Concerning my belief of God, I submit wholly to the Church, and believe as I have been informed out of the Athanasian Creed,[6] that the Father is incomprehensible, the Son incomprehensible, and the Holy Ghost incomprehensible; and that there are not three, but one incomprehensible God. Wherefore if any man can prove (as I do verily believe he cannot) that God is not incomprehensible, he must of necessity be more knowing than the whole Church, however he must needs dissent from the Church.

But perchance your author may say I raise new and prejudicial opinions in saying that matter is eternal. I answer, the Holy Writ does not mention matter to be created, but only particular [142] creatures, as this visible world, with all its parts, as the history or description of the creation of the world in Genesis plainly shows. For God said, "let it be light, and there was light; let there be a firmament in the midst of the waters, and let it divide the waters from the waters; and let the waters under the heaven be gathered together unto one place; and let the dry land appear; and let the earth bring forth grass, the herb yielding seed, and the fruit-tree yielding fruits after his kind; and let there be lights in the firmament of the heaven, to divide the day from the night, etc."[7] Which proves that all creatures and figures were made and produced out of that rude and desolate heap or chaos which the Scripture mentions, which is nothing else but matter, by the powerful word and command of God, executed by his eternal servant, nature, as I have heretofore declared it in a letter I sent you in the beginning concerning infinite nature.[8]

But lest I seem to encroach too much upon divinity, I submit this interpretation to the Church. However, I think it not against the ground of our faith; for I am so far from maintaining anything either against church or state, as I am submitting to both in all duty, and shall do so as long as I live, and rest, Madam,

Your faithful friend and servant.

---

6. The Athanasian Creed was a statement of religious belief in the doctrine of the Trinity that was included in the Church of England's *Book of Common Prayer*.

7. Cavendish paraphrases Gen. 1:3–15.

8. See in this volume, section 1, letter 3.

[143]   **4.**

Madam,

Since your worthy and learned author is pleased to mention that an "ample experience both of men and things does enlarge our understanding,"[9] I have taken occasion hence to enquire how a man's understanding may be increased or enlarged.

The understanding must either be in parts, or it must be individable as one; if in parts, then there must be so many understandings as there are things understood; but if individable, and but one understanding, then it must dilate itself upon so many several objects. I for my part assent to the first, that understanding increases by parts and not by dilation, which dilation must needs follow if the mind or understanding of man be indivisible and without parts; but if the mind or soul be individable, then I would fain know how understanding, imagination, conception, memory, remembrance, and the like, can be in the mind? You will say, perhaps, they are so many faculties or properties of the incorporeal mind, but I hope you do not intend to make the mind or soul a deity, with so many attributes.

Wherefore, in my opinion, it is safer to say that the mind is composed of several active parts; but of the parts I have treated in my philosophy, where you will find that all the several parts of nature are living and knowing, and that there is no part that has not life and knowledge, being all composed [144] of rational and sensitive matter, which is the life and soul of nature; and that nature, being material, is composable and dividable, which is the cause of so many several creatures, where every creature is a part of nature, and these infinite parts or creatures are nature herself; for though nature is a self-moving substance, and by self-motion divides and composes herself several manners or ways into several forms and figures, yet being a knowing as well as a living substance, she knows how to order her parts and actions wisely; for as she has an infinite body or substance, so she has an infinite life and knowledge; and as she has an infinite life and knowledge, so she has an infinite wisdom.

But mistake me not, Madam, I do not mean an infinite divine wisdom, but an infinite natural wisdom, given her by the infinite bounty of the omnipotent God; but yet this infinite wisdom, life, and knowledge in nature makes but one infinite. And as nature has degrees of matter, so she has also degrees and variety

---

9. More, *Antidote*, book 2, chapter 2, section 1.

of corporeal motions; for some parts of matter are self-moving and some are moved by these self-moving parts of matter; and all these parts, both the moving and moved, are so intermixed that none is without the other, no not in any the least creature or part of nature we can conceive; for there is no creature or part of nature but has a commixture of those mentioned parts of animate and inanimate matter, and all the motions are so ordered by nature's wisdom as not any thing in nature can be otherwise, unless by a supernatural command and power of God; for no part of corporeal matter and motion can either perish or but rest; one part may cause another part to alter its motion, but not to quit [145] motion, no more than one part of matter can annihilate or destroy another; and therefore matter is not merely passive, but always active, by reason of the thorough mixture of animate and inanimate matter; for although the animate matter is only active in its nature, and the inanimate passive, yet because they are so closely united and mixed together that they make but one body, the parts of the animate or self-moving matter do bear up and cause the inanimate parts to move and work with them; and thus there is an activity in all parts of matter moving and working as one body, without any fixation or rest, for all is moveable, moving, and moved.

All which, Madam, if it were well observed, there would not be so many strange opinions concerning nature and her actions, making the purest and subtlest part of matter immaterial or incorporeal, which is as much as to extend her beyond nature, and to rack her quite to nothing. But I fear the opinion of immaterial substances in nature will at last bring in again the heathen religion, and make us believe [in] a Pan, Bacchus, Ceres, Venus, and the like, so as we may become worshippers of groves and shadows, beans and onions, as our forefathers.

I say not this as if I would ascribe any worship to nature or make her a deity, for she is only a servant to God, and so are all her parts or creatures, which parts or creatures, although they are transformed, yet cannot be annihilated, except nature herself be annihilated, which may be whensoever the great God pleases; for her existence and resolution, or total destruction, depends upon God's will and decree, whom she fears, adores, admires, praises, and prays unto, as being her God and master; and as she adores [146] God, so do all her parts and creatures, and amongst the rest, man, so that there is no atheist in infinite nature, at least not in the opinion of, Madam,

Your faithful friend and servant.

5.

Madam,

I cannot well conceive what your author means by the "common laws of nature";[10] but if you desire my opinion how many laws nature has, and what they are, I say nature has but one law, which is a wise law, *viz.*, to keep infinite matter in order, and to keep so much peace, as not to disturb the foundation of her government. For though nature's actions are various, and so many times opposite, which would seem to make wars between several parts, yet those active parts, being united into one infinite body, cannot break nature's general peace; for that which man names "war," "sickness," "sleep," "death," and the like, are but various particular actions of the only matter; not, as your author imagines, in a confusion, like bullets, or such like things juggled together in a man's hat, but very orderly and methodical. And the playing motions of nature are the actions of art, but her serious actions are the actions of production, generation, and [147] transformation in several kinds, sorts, and particulars of her creatures, as also the action of ruling and governing these her several active parts.

Concerning the pre-eminence and prerogative of man, whom your author calls "the flower and chief of all the products of nature upon this globe of the earth,"[11] I answer that man cannot well be judged of himself, because he is a party, and so may be partial; but if we observe well, we shall find that the elemental creatures are as excellent as man, and as able to be a friend or foe to man, as man to them, and so the rest of all creatures; so that I cannot perceive more abilities in man than in the rest of natural creatures; for though he can build a stately house, yet he cannot make a honey-comb; and though he can plant a slip,[12] yet he cannot make a tree; though he can make a sword or knife, yet he cannot make the metal. And as man makes use of other creatures, so other creatures make use of man, as far as he is good for anything. But man is not so useful to his neighbor or fellow-creatures as his neighbor or fellow-creatures to him, being not so profitable for use, as apt to make spoil. And so leaving him, I rest, Madam,

Your faithful friend and servant.

---

10. More, *Antidote*, book 2, chapter 2, section 13.

11. More, *Antidote*, book 2, chapter 3, section 3.

12. Slip: twig or cutting from a tree.

Madam,

Your author demands "whether there was ever any man, that was not a mortal, and whether there be any mortal that had not a beginning?"[13] Truly, if nature be eternal, all the material figures which ever were, are, and can be, must be also eternal in nature, for the figures cannot be annihilated, unless nature be destroyed; and although a creature is dissolved and transformed into numerous different figures, yet all these several figures remain still in those parts of matter, whereof that creature was made, for matter never changes, but is always one and the same, and figure is nothing else but matter transposed or transformed by motion several modes or ways. But if you conceive matter to be one thing, figure another, and motion a third, several, distinct, and dividable from each other, it will produce gross errors, for matter, motion, and figure are but one thing.

And as for that common question, whether the egg was before the chick or the chick before the egg, it is but a threadbare argument which proves nothing, for there is no such thing as first in eternity, neither does time make productions or generations, but matter; and whatsoever matter can produce or generate was in matter before it was produced; wherefore the question is whether matter, which is nature, had a beginning or not? I say not; for put the case, the figures of earth, air, water, and fire, [149] light and colors, heat and cold, animals, vegetables, and minerals, etc. were not produced from all eternity, yet those figures have nevertheless been in matter, which is nature, from all eternity, for these mentioned creatures are only made by the corporeal motions of matter, transforming matter into such several figures.

Neither can there be any perishing or dying in nature, for that which man calls so is only an alteration of figure. And as all other productions are but a change of matter's sensitive motions, so all irregular and extravagant opinions are nothing but a change of matter's rational motions; only productions by rational motions are interior, and those by sensitive motions exterior. For the natural mind is not less material than the body, only the matter of the mind is much purer and subtler than the matter of the body. And thus there is nothing in nature but what is material; but he that thinks it absurd to say the world is composed of mere self-moving matter may consider that it is more absurd to believe immaterial substances or spirits in nature, as also a spirit of nature,[14] which is

---

13. More, *Antidote*, book 3, chapter 15, section 3.

14. On More's "spirit of nature," see the Introduction (p. xxxvi).

the vicarious power of God upon matter. For why should it not be as probable that God did give matter a self-moving power to herself as to have made another creature to govern her? For nature is not a babe or child, to need such a spiritual nurse to teach her to go or to move; neither is she so young a lady as to have need of a governess, for surely she can govern herself; she needs not a guardian for fear she should run away [150] with a younger brother,[15] or one that cannot make her a jointure.[16] But leaving those strange opinions to the fancies of their authors, I'll add no more, but that I am, Madam,

Your faithful friend and servant.

7.

Madam,

Your author being very earnest in arguing against those that maintain the opinion of matter being self-moving, amongst the rest of his arguments brings in this: "Suppose," says he, "matter could move itself, would mere matter with self-motion amount to that admirable wise contrivance of things which we see in the world?—All the evasion I can imagine our adversaries may use here, will be this: that matter is capable of sense; and the finest and most subtle of the most refined sense; and consequently of imagination too, yea happily of reason and understanding."[17]

I answer, it is very probable that not only all the matter in the world or universe has sense, but also reason; and that the sensitive part of matter is the builder, and the rational the designer, whereof I have spoken of before, and you may find more of it in my book of philosophy.[18] "But," says your author, "let us see if all their heads laid [151] together can contrive the anatomical fabric of any

---

15. According to laws of primogeniture at that time, only the oldest son could inherit a father's wealth; marrying a younger brother rather than the oldest son in a family would mean less financial stability for a woman.

16. That is, with a suitor with too little money to provide for her if he were to die after their marriage.

17. More, *Immortality*, book 1, chapter 12, sections 1 and 2.

18. See Cavendish, *Philosophical and Physical Opinions* (1663), part 1, chapter 3.

creature that lives?"[19] I answer, all parts of nature are not bound to have heads or tails; but if they have, surely they are wiser than many a man's. "I demand," says he, "has every one of these particles, that must have a hand in the framing of the body of an animal, the whole design of the work by the impress of some phantasm upon it? or as they have several offices,[20] so have they several parts of the design?" I answer, all the actions of self-moving matter are not impresses, nor is every part a hand-laborer, but every part unites by degrees into such or such a figure. Again, says he, "how is it conceivable that one particle of matter, or many together (there not existing yet in nature an animal) can have the idea impressed of that creature they are to frame?" I answer, all figures whatsoever have been, are, or can be in nature, are existent in nature. "How," says he, "can they in framing several parts confer notes? By what language or speech can they communicate their counsels one to another?" I answer, knowledge does not always require speech, for speech is an effect and not a cause, but knowledge is a cause and not an effect; and nature has infinite more ways to express knowledge than man can imagine.

"Wherefore," he concludes, "that they should mutually serve one another in such a design, is more impossible, than that so many men, blind and dumb from their nativity, should join their forces and wits together to build a castle, or carve a statue of such a creature, as none of them knew any more in several, than some one of the smallest parts thereof, but not the relation it bore to the whole."[21]

I answer, nature is neither blind nor dumb, nor any ways defective, but infinitely wise and knowing; for [152] blindness and dumbness are but effects of some of her particular actions, but there is no defect in self-moving matter, nor in her actions in general; and it is absurd to conceive the generality of wisdom according to an irregular effect or defect of a particular creature; for the general actions of nature are both life and knowledge, which are the architects of all creatures and know better how to frame all kinds and sorts of creatures than man can conceive; and the several parts of matter have a more easy way of communication than man's head has with his hand, or his hand with pen, ink, and paper, when he is going to write; which latter example will make you understand my opinion the better, if you do but compare the rational part of matter to the head, the sensitive to the hand, the inanimate to pen, ink, and paper, their action

---

19. More, *Immortality*, book 1, chapter 12, section 4; all subsequent quotations from More in this paragraph are from section 4.

20. Offices: duties; functions.

21. More, *Immortality*, book 1, chapter 12, section 4.

to writing, and their framed figures to those figures or letters which are written; in all which is a mutual agreement without noise or trouble.

But give me leave, Madam, to tell you that self-moving matter may sometimes err and move irregularly, and in some parts not move so strong, curious, or subtle at some times, as in other parts, for nature delights in variety. Nevertheless, she is more wise than any particular creature or part can conceive, which is the cause that man thinks nature's wise, subtle, and lively actions are as his own gross actions, conceiving them to be constrained and turbulent, not free and easy, as well as wise and knowing. Whereas nature's creating, generating, and producing actions are by an easy connection of parts to parts, without counterbuffs, jogs, and jolts, producing a particular figure by degrees, and in order and method, as human sense and reason may [153] well perceive. And why may not the sensitive and rational part of matter know better how to make a bee than a bee does how to make honey and wax? Or have a better communication between them than bees that fly several ways, meeting and joining to make their combs in their hives?

But pardon, Madam, for I think it a crime to compare the creating, generating, and producing corporeal life and wisdom of nature unto any particular creature, although every particular creature has their share, being a part of nature. Wherefore those, in my opinion, do grossly err that bind up the sensitive matter only to taste, touch, hearing, seeing, and smelling; as if the sensitive parts of nature had not more variety of actions than to make five senses; for we may well observe, in every creature there is difference of sense and reason according to the several modes of self-motion. For the sun, stars, earth, air, fire, water, plants, animals, minerals, although they have all sense and knowledge, yet they have not all sense and knowledge alike, because sense and knowledge moves not alike in every kind or sort of creature; but yet this does not cause a general ignorance, as to be altogether insensible or irrational, neither do the erroneous and irregular actions of sense and reason prove an annihilation of sense and reason; as for example, a man may become mad or a fool through the irregular motions of sense and reason, and yet have still the perception of sense and reason, only the alteration is caused through the alteration of the sensitive and rational corporeal motions or actions from regular to irregular; nevertheless he has perceptions, [154] thoughts, ideas, passions, and whatsoever is made by sensitive and rational matter; neither can perception be divided from motion, nor motion from matter, for all sensation is corporeal, and so is perception. I can add no more, but take my leave, and rest, Madam,

Your faithful friend and servant.

8.

Madam,

Your author is pleased to say that "matter is a principle purely passive, and no otherwise moved or modified, than as some other thing moves and modifies it, but cannot move itself at all; which is most demonstrable to them that contend for sense and perception in it. For if it had any such perception, it would, by virtue of its self-motion withdraw itself from under the knocks of hammers, or fury of the fire; or of its own accord approach to such things as are most agreeable to it, and pleasing, and that without the help of muscles, it being thus immediately endowed with a self-moving power."[22]

By his leave, Madam, I must tell you that I see no consequence in this argument. Because some parts of matter cannot withdraw themselves from the force and power of other parts, therefore they have neither sense, reason, nor perception. For put the case, a man should be [155] overpowered by some other men, truly he would be forced to suffer, and no immaterial spirits, I think, would assist him. The very same may be said of other creatures or parts of nature; for some may overpower others, as the fire, hammer, and hand do overpower a horse-shoe, which cannot prevail over so much odds of power and strength. And so likewise it is with sickness and health, life and death; for example, some corporeal motions in the body turning rebels, by moving contrary to the health of an animal creature, it must become sick; for not every particular creature has an absolute power, and not in single divided parts.

Indeed, to speak properly, there is no such thing as an absolute power in nature; for though nature has power to move itself, yet not beyond itself. But mistake me not, for I mean by an absolute power, not a circumscribed and limited, but an unlimited power, no ways bound or confined, but absolutely or every way infinite, and there is not anything that has such an absolute power but God alone. Neither can nature be undividable, being corporeal or material; nor rest from motion, being naturally self-moving and in a perpetual motion. Wherefore though matter is self-moving and very wise (although your author denies it, calling those "fools" that maintain this opinion[23]), yet it cannot go beyond the rules of its nature, no more than any art can go beyond its rules and principles.

---

22. More, *Immortality*, book 2, chapter 1, section 3.

23. More, *Antidote*, appendix, chapter 3, section 10.

And as for what your author says, that everything would approach to that which is agreeable and pleasant, I think I need no demonstration to prove it; for we may plainly see it in all effects of nature, that there is [156] sympathy and antipathy, and what is this else but approaching to things agreeable and pleasant, and withdrawing itself from things disagreeable, and hurtful or offensive? But of this subject I shall discourse more hereafter, wherefore I finish here, and rest, Madam,

> Your faithful friend and servant.

9.

Madam,

Your author's opinion is that "matter being once actually divided as far as it possibly can, it is a perfect contradiction it should be divided any further."[24] I answer, though nature is infinite, yet her actions are not all dilative nor separative, but some divide and some compose, some dilate and some contract, which causes a mean between nature's actions or motions.

Next your author says that "as infinite greatness has no figure, so infinite littleness has none also."[25] I answer, whatsoever has a body has a figure; for it is impossible that substance or body, and figure, should be separated from each other, but wheresoever is body or substance, there is also figure, and if there be an infinite substance, there must also be an infinite figure, although not a certain determined or circumscribed figure, for such a figure belongs only to finite particulars; [157] and therefore I am of your author's mind that it is a contradiction to say an infinite cube or triangle, for a cube or triangle is a perfect circumscribed figure, having a certain compass and circumference, be it never so great or little; wherefore to say an infinite cube, would be as much as to say a finite infinite. [. . .]

But mistake me not, when I speak of circumscribed [158] and finite single parts; for I do not mean that each part does subsist single and by itself, there being no such thing as an absolute single part in nature, but infinite matter being by self-motion divided into an infinite number of parts, all these parts have so

---

24. More, *Immortality*, preface, section 3.
25. More, *Immortality*, preface, section 3.

near a relation to each other and to the infinite whole that one cannot subsist without the other; for the infinite parts in number do make the infinite whole, and the infinite whole consists in the infinite number of parts; wherefore it is only their figures which make a difference between them; for each part having its proper figure different from the other, which is circumscribed and limited, it is called a finite single part; and such a part cannot be said [to be] infinitely dividable, for infinite composition and division belong only to the infinite body of nature, which being infinite in substance may also be infinitely divided, but not a finite and single part. Besides, infinite composition does hinder the infinite division, and infinite division hinders the infinite composition; so that one part cannot be either infinitely composed or infinitely divided; and it is one thing to be dividable and another to be divided.

And thus, when your author mentions in another place that "if a body be dividable into infinite parts, it has an infinite number of extended parts,"[26] if by extension he means corporeal dimension, I am of his opinion; for there is no part, be it ever so little in nature, but is material; and if material, it has a body; and if a body, it must needs have a bodily dimension; and so every part will be an extended part. But since there is no part but is finite in itself, it cannot be divisible into infinite parts; neither can any part be infinitely dilated or contracted; for as composition [159] and division do hinder and obstruct each other from running into infinite, so does dilation hinder the infinite contraction, and contraction the infinite dilation, which, as I said before, causes a mean between nature's actions; nevertheless, there are infinite dilations and contractions in nature, because there are infinite contracted and dilated parts, and so are infinite divisions because there are infinite divided parts; but contraction, dilation, extension, composition, division, and the like are only nature's several actions; and as there can be no single part in nature that is infinite, so there can neither be any single infinite action.

But as for matter, motion, and figure, those are individable and inseparable, and make but one body or substance; for it is as impossible to divide them, as impossible it is to your author to separate the essential proprieties which he gives from an immortal spirit; so is likewise matter, space, place, and duration. For parts, motion, figure, place, and duration, are but one infinite body; only the infinite parts are the infinite divisions of the infinite body, and the infinite body is a composition of the infinite parts; but figure, place, and body are all one, and so is time and duration, except you will call time the division of duration,

---

26. More, *Antidote*, book 1, chapter 4, section 2 (misidentified by Cavendish as book 2).

and duration the composition of time; but infinite time and infinite duration is all one in nature. And thus nature's principal motions and actions are dividing, composing, and disposing or ordering, according to her natural wisdom, by the omnipotent God's leave and permission. [. . .]

[160] I take my leave, and rest, Madam,

Your faithful friend and servant.

10.

Madam,

Your author, in his arguments against motion being a principle of nature, endeavors to prove that beauty, color, symmetry, and the like, in plants as well as in other creatures, are no result from the mere motion of the matter; and forming this [161] objection, "it may be said," says he, "that the regular motion of the matter made the first plant of every kind; but we demand, what regulated the motion of it, so as to guide it, to form itself into such a state?"[27]

I answer, the wisdom of nature or infinite matter did order its own actions so as to form those her parts into such an exact and beautiful figure, as such a tree, or such a flower, or such a fruit, and the like; and some of her parts are pleased and delighted with other parts, but some of her parts are afraid or have an aversion to other parts; and hence is like and dislike, or sympathy and antipathy, hate and love, according as nature, which is infinite self-moving matter, pleases to move; for though natural wisdom is dividable into parts, yet these parts are united in one infinite body, and make but one being in itself, like as the several parts of a man make up but one perfect man; for though a man may be wise in several causes[28] or actions, yet it is but one wisdom; and though a judge may show justice in several causes, yet it is but one justice; for wisdom and justice, though they be practiced in several causes, yet it is but one wisdom and one justice; and so, all the parts of a man's body, although they move differently, yet are they but one man's bodily actions. Just as a man, if he carve or cut out by art several statues or draw several pictures, those statues or pictures are but that one man's work.

---

27. More, *Antidote*, appendix, chapter 11, section 4.
28. Causes: legal cases.

The like may be said of nature's motions and figures; all which are but one self-active or self-moving material nature. But wise nature's ground or fundamental actions are very regular, as you may observe in the several and distinct kinds, sorts, and particulars of her creatures, and in their distinct [162] proprieties, qualities, and faculties, belonging not only to each kind and sort, but to each particular creature; and since man is not able to know perfectly all those proprieties which belong to animals, much less will he be able to know and judge of those that are in vegetables, minerals, and elements; and yet these creatures, for anything man knows, may be as knowing, understanding, and wise as he; and each as knowing of its kind or sort as man is of his. But the mixture of ignorance and knowledge in all creatures proceeds from thence, that they are but parts; and there is no better proof that the mind of man is dividable than that it is not perfectly knowing; nor no better proof that it is composable than that it knows so much.

But all minds are not alike, but some are more composed than others, which is the cause [that] some know more than others; for if the mind in all men were alike, all men would have the same imaginations, fancies, conceptions, memories, remembrances, passions, affections, understanding, and so forth. The same may be said of their bodies; for if all men's sensitive parts were as one, and not dividable and composable, all their faculties, proprieties, constitutions, complexions, appetites would be the same in every man without any difference; but human sense and reason do well perceive, that neither the mind, life, nor body are as one piece, without division and composition.

Concerning the divine soul, I do not treat of it; only this I may say, that all are not devout alike, nor those which are, are not at all times alike devout. But to conclude: some of our modern philosophers think they do God good service when they endeavor to prove nature, as God's good servant, to [163] be stupid, ignorant, foolish, and mad, or anything rather than wise, and yet they believe themselves wise, as if they were no part of nature; but I cannot imagine any reason why they should rail on her, except nature had not given them as great a share or portion as she has given to others; for children in this case do often rail at their parents for loving their brothers and sisters more than themselves. However, nature can do more than any of her creatures. And if man can paint, embroider, carve, engrave curiously;[29] why may not nature have more ingenuity,

---

29. Curiously: skillfully, excellently.

wit, and wisdom than any of her particular creatures? The same may be said of her government. And so leaving wise nature, I rest, Madam,

Your faithful friend and servant.

11.

Madam,

To your author's argument that "if motion belong naturally to matter, matter being uniform, it must be alike moved in every part or particle imaginable of it, by reason this motion being natural and essential to matter, is alike every way,"[30] I answer that this is no more necessary than that the several actions of one body or of one part of a body should be alike; for though matter is one and the same in its nature [164] and never changes, yet the motions are various, which motions are the several actions of one and the same natural matter; and this is the cause of so many several creatures, for self-moving matter by its self-moving power can act several ways, modes, or manners; and had not any natural matter a self-acting power, there could not be any variety in nature; for nature knows of no rest, there being no such thing as rest in nature; but she is in a perpetual motion, I mean self-motion, given her from God.

Neither do I think it atheistical (as your author deems[31]) to maintain this opinion of self-motion, as long as I do not deny the omnipotency of God; but I should rather think it irreligious to make so many several creatures as immaterial spirits, like so many several deities, to rule and govern nature and all material substances in nature; for what atheism does there lie in saying that natural matter is naturally moving and wise in herself? Does this oppose the omnipotency and infinite wisdom of God? It rather proves and confirms it; for all nature's free power of moving and wisdom is a gift of God and proceeds from him; but I must confess, it destroys the power of immaterial substances, for nature will not be ruled nor governed by them, and to be against natural immaterial substance I think is no atheism, except we make them deities; neither is it atheism to contradict the opinion of those that believe such natural incorporeal spirits, unless man make himself a god.

---

30. More, *Antidote*, book 2, chapter 1, section 2.
31. More, *Antidote*, book 2, chapter 1, section 5.

But although nature is wise, as I said before, and acts methodically, yet the variety of motions is the cause of so many irregularities in nature, as also the cause of irregular opinions; for all opinions are made by self-moving matter's [165] motions, or (which is all one) by corporeal self-motion, and some in their opinions do conceive nature according to the measure of themselves, as that nature can nor could not do more than they think; nay, some believe they can do as much as nature does; which opinions, whether they be probable or regular, I'll let any man judge; adding only this, that to human sense and reason it appears plainly that as God has given nature a power to act freely, so he does approve of her actions, being wise and methodical in all her several productions, generations, transformations, and designs. And so I conclude for the present, only subscribe myself as really I am, Madam,

Your faithful friend and servant.

12.

Madam,

I am of your author's opinion concerning self-activity or self-motion, that "what is active of itself, can no more cease to be active than to be."[32] And I have been always of this opinion, even from the first beginning of my conceptions in natural philosophy, as you may see in my first treatise of natural philosophy, which I put forth eleven years since, where I say that self-moving matter is in a perpetual motion;[33] but your author [166] endeavors from thence to conclude that "matter is not self-active, because it is reducible to rest." To which I answer that there is no such thing as rest in nature.

Nor do I say that all sorts of motions are subject to our senses, for those that are subject to our sensitive perceptions are but gross motions in comparison to those that are not subject to our exterior senses. As for example, we see some

---

32. More, *Immortality*, book 1, chapter 7, section 1. More actually refers to what is "simply active of itself."

33. That is, Cavendish's 1653 *Philosophicall Fancies* (the precursor to the 1655 and 1663 editions of *Philosophical and Physical Opinions*). In this book, Cavendish does not explicitly say that self-moving matter is in perpetual motion; however, in the chapter "Of Life," she says that life is the "extract" (essence) of "common matter," and in the chapter "Of Decay," she says that the "nature of life" is "a perpetual motion."

bodies dilate, others consume, others corrupt; yet we do not see how they dilate, nor how they consume, nor how they corrupt. Also we see some bodies contract, some attract, some condense, some consist,[34] etc., yet we do not see their contracting, attracting, condensing, consisting, or retenting motions; and yet we cannot say they are not corporeal motions, because not subject to our exterior senses; for if there were not contracting, attracting, retenting, or consistent corporeal self-motions, it had been impossible that any creature could have been composed into one united figure, much less stayed and continued in the same figure without a general alteration.

But your author says, "if matter, as matter, had motion, nothing would hold together, but flints, adamant,[35] brass, iron, yea, this whole earth, would suddenly melt into a thinner substance than the subtle air, or rather it never had been condensated together to this consistency we find it."[36] But I would ask him what reason he can give that corporeal self-motion should make all matter rare and fluid, unless he believe there is but one kind of motion in nature, but this, human sense and reason will contradict; for we may observe there are infinite changes of motion, and there is more variety and curiosity in corporeal motions than any one single creature can imagine, [167] much less know; but I suppose he conceives all corporeal matter to be gross, and that not any corporeal motion can be subtle, penetrating, contracting, and dilating; and that whatsoever is penetrating, contracting, and dilating is individable.

But by his leave, Madam, this does not follow; for though there be gross degrees of matter and strong degrees of corporeal motions, yet there are also pure and subtle degrees of matter and motions; to wit, that degree of matter which I name sensitive and rational matter, which is natural life and knowledge, as sensitive life and rational knowledge. [. . .]

[168] And thus I conclude, and rest constantly, Madam,

Your faithful friend and servant.

---

34. Consist: hold together.
35. Adamant: a very hard rock.
36. More, *Immortality*, book 1, chapter 7, section 1.

13.

Madam,

That matter is incapable of sense, your author proves by the example of dead carcasses. For, says he, "motion and sense being really one and the same thing, it must needs follow that where there is motion, there is also sense and perception; but on the contrary, there is reaction in dead carcasses, and yet no sense."[37] I answer shortly that it is no consequence [that] because there is no animal sense nor exterior perceptible local motion in a dead carcass, therefore there is no sense at all in it; for though it has not animal sense, yet it may nevertheless have sense according to the nature of that figure into which it did change from being an animal.

Also he says, "if matter have sense, it will follow, that upon reaction all shall have the like; and that a bell while it is ringing, and a bow while it is bent, and every jack-in-a-box that school-boys play with, shall be [169] living animals."[38] I answer, it is true, if reaction made sense; but reaction does not make sense, but sense makes reaction, and though the ball has not an animal knowledge, yet it may have a mineral life and knowledge, and the bow and the jack-in-a-box a vegetable knowledge; the shape and form of the bell, bow, and jack-in-a-box is artificial; nevertheless each in its own kind may have as much knowledge as an animal in his kind; only they are different according to the different pro-prieties of their figures. And who can prove the contrary that they have not? For certainly man cannot prove what he cannot know; but man's nature is so, that knowing but little of other creatures, he presently judges there is no more knowledge in nature than what man, at least animals have; and confines all sense only to animal sense, and all knowledge to animal knowledge.

Again, says your author, "that matter is utterly uncapable of such operations as we find in ourselves, and that therefore there is something in us immaterial or incorporeal; for we find in ourselves that one and the same thing both hears, and sees, and tastes, and perceives all the variety of objects that nature manifests unto us."[39] I answer, that is the reason there is but one matter, and that all nat-ural perception is made by the animate part of matter; but although there is but one matter in nature, yet there are several parts or degrees, and consequently

---

37. More, *Immortality*, book 2, chapter 2, section 1.

38. More, *Immortality*, book 2, chapter 2, section 1. More actually says they "shall be living ani-mals, or sensitive creatures."

39. More, *Immortality*, book 2, chapter 2, section 2.

several actions of that only matter, which causes such a variety of perceptions, both sensitive and rational. The sensitive perception is made by the sensitive corporeal motions, copying out the figures of foreign objects in the sensitive organs of the sentient; and if those sensitive motions do [170] pattern out foreign objects in each sensitive organ alike at one and the same time, then we hear, see, taste, touch, and smell at one and the same time. But thoughts and passions, as imagination, conception, fancy, memory, love, hate, fear, joy, and the like, are made by the rational corporeal motions in their own degree of matter, to wit, the rational. And thus all perception is made by one and the same matter, through the variety of its actions or motions, making various and several figures, both sensitive and rational.

But all this variety in sense and reason, or of sensitive and rational perceptions, is not made by parts pressing upon parts, but by changing their own parts of matter into several figures by the power of self-motion. For example, I see a man or beast; that man or beast does not touch my eye in the least, neither in itself nor by pressing the adjoining parts; but the sensitive corporeal motions straight upon the sight of the man or beast make the like figure in the sensitive organ, the eye, and in the eye's own substance or matter, to wit, the rational and inanimate, for they are all mixed together.

But this is to be observed, that the rational matter can and does move in its own substance, as being the purest and subtlest degree of matter; but the sensitive, being not so pure and subtle, moves always with the inanimate matter, and so the perceptive figures which the rational matter or rational corporeal motions make are made in their own degree of matter; but those figures which the sensitive [matter] patterns out are made in the organs or parts of the sentient body proper to such or such a sense or perception. As in an animal creature, the perception of sight [171] is made by the sensitive corporeal motions in the eye; the perception of hearing, in the ear; and so forth. [. . .]

[173] But men not considering or believing there might be such a degree of only matter, namely rational, it has made them err in their judgments. Nevertheless there is a difference between sensitive and rational parts and motions, and yet they are agreeable most commonly in their actions, though not always. Also the rational can make such figures as the sensitive cannot, by reason the rational has a greater power and subtler faculty in making variety than the sensitive; for the sensitive is bound to move with the inanimate, but the rational moves only in its own parts; for though the sensitive and rational oftentimes cause each other to move, yet they are not of one and the same degree of matter, nor have they the same motions.

And this rational matter is the cause of all notions, conceptions, imaginations, deliberation, determination, memory, and anything else that belongs to the mind; for this matter is the mind of nature, and so, being dividable, the mind of all creatures, as the sensitive is the life; and it can move, as I said, more subtly and more variously than the sensitive, and make such figures as the sensitive cannot, without outward examples and objects.

But all diversity comes by change of motion, and motions are as sympathetical and agreeable as antipathetical and disagreeing. And though nature's artificial motions, which are her playing motions, are sometimes extravagant, yet in her fundamental actions there is no extravagancy, as we may observe [174] by her exact rules in the various generations, the distinct kinds and sorts, the several exact measures, times, proportions, and motions of all her creatures, in all which her wisdom is well expressed, and in the variety her wise pleasure [is expressed]. To which I leave her, and rest, Madam,

Your faithful friend and servant.

14.

Madam,

"If there be any sense and perception in matter," says your author, "it must needs be motion or reaction of one part of matter against another; and that all diversity of sense and perception does necessarily arise from the diversity of the magnitude, figure, posture, vigor, and direction of motion in parts of matter. In which variety of perceptions, matter has none but such as are impressed by corporeal motions, that is to say, that are perceptions of some actions, or modificated impressions of parts of matter bearing one against another."[40]

I have declared, Madam, my opinion concerning perception in my former letters,[41] that all perception is not impression and reaction, like as a seal is printed on wax. For example, the corporeal rational motions in the mind do not print,

---

40. More, *Immortality*, book 2, chapter 1, sections 1, 6, 7. From section 1, Cavendish refers to More's Axiom 20, "Motion or reaction of one part of the matter against another, or at least a due continuance thereof, is really one and the same with sense and perception, if there be any sense or perception in matter." The other parts of her quotation are More's Axiom 22 (from section 6) and Axiom 23 (from section 7).

41. See in this volume, section 1, letters 4, 5, 18, 22, 24, 25, and 37.

but move figuratively; but the sensitive motions do carve, print, engrave, and, as it were [175] pencil out, as also move figuratively in productions, and do often take patterns from the rational figures, as the rational motions make figures according to the sensitive patterns. But the rational can move without patterns, and so the sensitive. For surely, were a man born blind, deaf, dumb, and had a numb palsy[42] in his exterior parts, the sensitive and rational motions would nevertheless move both in body and mind according to the nature of his figure; for though no copies were taken from outward objects, yet he would have thoughts, passions, appetites, and the like; and though he could not see exterior objects, nor hear exterior sounds, yet no question but he would see and hear interiously after the manner of dreams, only they might not be anything like to what is perceivable by man in the world; but if he sees not the sunlight, yet he would see something equivalent to it; and if he hears not such a thing as words, yet he would hear something equivalent to words; for it is impossible that his sensitive and rational faculties should be lost for want of an ear or an eye; so that perception may be without exterior objects, or marks, or patterns. For although the sensitive motions do usually pattern out the figures of exterior objects, yet that does not prove, but they can make interior figures without such objects.

Wherefore perception is not always reaction, neither is perception and reaction really one thing; for though perception and action is one and the same, yet not always reaction; but did perception proceed from the reaction of outward objects, a blind and deaf man would not so much as dream; for he would have no interior motion in the head, having no other exterior sense but touch, [176] which, if the body was troubled with a painful disease, he would neither be sensible of, but to feel pain, and interiously feel nothing but hunger and fullness; and his mind would be as irrational as some imagine vegetables and minerals are. To which opinion I leave them, and rest, Madam,

Your faithful friend and servant.

---

42. Palsy: paralysis or weakness.

15.

Madam,

Your author is pleased, in mirth, and to disgrace the opinion of those which hold that perception is made by figuring, to bring in this following example: "Suppose," says he, "one particle should shape itself into a George on horseback with a lance in his hand, and another into an enchanted castle; this George on the horseback must run against the castle, to make the castle receive his impress and similitude. But what then? Truly the encounter will be very unfortunate, for Saint George indeed may easily break his lance, but it is impossible that he should by jostling against the particle in the form of a castle, convey the entire shape of himself and his horse thereby, such as we find ourselves able to imagine of a man on horseback; which is a truth as demonstrable as any theorem in mathematics."[43]

I answer, [177] first, that there is no particle single and alone by itself. Next, I say it is more easy for the rational matter to put itself into such figures and to make such encounters than for an immaterial mind or substance to imagine it; for no imagination can be without figure, and how should an immaterial created substance present such figures but by making them either in itself or upon matter? For Saint George and the castle are figures, and their encounters are real fighting actions, and how such figures and actions can be in the mind or memory, and yet not be, is impossible to conceive; for, as I said, those figures and actions must be either in the incorporeal mind or in the corporeal parts of matter; and if the figures and motions may be in an incorporeal substance, much more is it probable for them to be in a corporeal; nay, if the figures and their actions can be in gross corporeal matter, why should they not be in the purest part of matter, which is the rational matter? And as for being made known to the whole body and every part thereof, it is not necessary, no more than it is necessary that the private actions of every man or family should be made known to the whole kingdom, or town, or parish.

But my opinion of self-corporeal motion and perception may be as demonstrable as that of immaterial natural spirits, which, in my mind, is not demonstrable at all, by reason it is not corporeal or material. For how can that be naturally demonstrable which naturally is nothing? But your author believes the mind or rational soul to be individable, and therefore concludes that the parts of

---

43. More, *Immortality*, book 2, chapter 6, section 6. More refers to the medieval legend of Saint George, a knight on horseback who slayed a dragon.

the same matter, although at great distance, must of necessity know each particular act of each several part; but that is not [178] necessary; for if there were not ignorance through the division of parts, every man and other creatures would know alike; and there is no better proof that matter or any particular creature in nature is not governed by a created immaterial spirit than that knowledge is in parts; for the hand does not know what pain the head feels, which certainly it would do, if the mind were not dividable into parts, but an individable substance.

But this is well to be observed, that some parts in some actions agree generally in one body, and some not; as for example, temperance and appetite do not agree; for the corporeal actions of appetite desire to join with the corporeal actions of such or such other parts, but the corporeal actions of temperance do hinder and forbid it; whereupon there is a faction amongst the several parts. For example, a man desires to be drunk with wine; this desire is made by such corporeal actions as make appetite; the rational corporeal motions or actions which make temperance oppose those that make appetite, and that sort of actions which has the better carries it, the hand and other parts of the body obeying the strongest side; and if there be no wine to satisfy the appetite, yet many times the appetite continues; that is, the parts continue in the same motions that make such an appetite; but if the appetite does not continue, then those parts have changed their motions; or when by drinking, the appetite is satisfied and ceases, then those parts that made the appetite have altered their former motions.

But oftentimes the rational corporeal motions may so agree with the sensitive as there may be no opposition or crossing at all, but a sympathetical mutual agreement [179] between them, at least an approvement;[44] so that the rational may approve what the sensitive covet or desire. Also some motions of the rational, as also of the sensitive matter, may disagree amongst themselves, as we see that a man will often have a divided mind; for he will love and hate the same thing, desire and not desire one and the same thing, as to be in heaven and yet to be in the world.

Moreover, this is to be observed, that all rational perceptions or cogitations are not so perspicuous and clear as if they were mathematical demonstrations, but there is some obscurity more or less in them; at least, they are not so well perceivable without comparing several figures together, which proves they are not made by an individable, immaterial spirit, but by dividable corporeal parts. As for example, man writes oftentimes false, and seldom so exact, but he is

---

44. Approvement: approval.

forced to mend his hand[45] and correct his opinions, and sometimes quite to alter them, according as the figures continue or are dissolved and altered by change of motion; and according as the actions are quick or slow in these alterations, the human mind is settled or wavering; and as figures are made or dissolved and transformed, opinions, conceptions, imaginations, understanding, and the like, are more or less. And according as these figures last, so is constancy, memory, or forgetfulness, and as those figures are repeated, so is remembrance; but sometimes they are so constant and permanent, as they last as long as the figure of the body, and sometimes it happens not once in an age that the like figures are repeated, and sometimes they are repeated every moment. As for example, a man remembers or calls to mind the figure of another man, his friend, with all his qualities, [180] dispositions, actions, proprieties, and the like, several times in an hour, and sometimes not once in a year, and so often as he remembers him, as often is the figure of that man repeated; and as oft as he forgets him, so often is his figure dissolved.

But some imagine the rational motions to be so gross as the trotting of a horse, and that all the motions of animate matter are as rude and coarse as renting[46] or tearing asunder, or that all impressions must needs make dents or creases. But as nature has degrees of corporeal motions, matter and motion being but one substance; and it is absurd to judge of the interior motions of self-moving matter by artificial or exterior gross motions, as that all motions must be like the tearing of a sheet of paper, or that the printing and patterning of several figures of rational and sensitive matter must be like the printing of books; nay, all artificial printings are not so hard as to make dents and impresses; witness writing, painting, and the like; for they do not disturb the ground whereon the letters are written or the picture drawn, and so the curious actions of the purest rational matter are neither rude nor rough; but although this matter is so subtle and pure as not subject to exterior human senses and organs, yet certainly it is dividable, not only in several creatures, but in the several parts of one and the same creature, as well as the sensitive, which is the life of nature, as the other is the soul; not the divine, but natural soul; neither is this soul immaterial, but corporeal; not composed of rags and shreds, but it is the purest, simplest, and subtlest matter in nature.

But to conclude, I desire you to remember, Madam, that this [181] rational and sensitive matter in one united and finite figure or particular creature has both common and particular actions; for as there are several kinds and sorts of

---

45. Hand: handwriting.
46. Renting: rending, ripping.

creatures and particulars in every kind and sort; so the like for the actions of the rational and sensitive matter in one particular creature. Also it is to be noted that the parts of rational matter can more suddenly give and take intelligence to and from each other than the sensitive; nevertheless, all parts in nature, at least adjoining parts, have intelligence[47] between each other, more or less, because all parts make but one body; for it is not with the parts of matter as with several constables in several hundreds,[48] or several parishes, which are a great way distant from each other, but they may be as close as the combs of bees, and yet as partable and as active as bees. But concerning the intelligence of nature's parts, I have sufficiently spoken in other places;[49] and so I'll add no more, but that I unfeignedly remain, Madam,

Your faithful friend and servant.

[183]   **17.**

Madam,

Outward objects, as I have told you before, do not make sense and reason, but sense and reason do perceive and judge of outward objects; for the sun does not make sight, nor does sight make light; but sense and reason in a man or any other creature do perceive and know there are such objects as sun and light, or whatsoever objects are presented to them. Neither does dumbness, deafness, blindness, etc. cause an insensibility, but sense through irregular actions causes them; I say through irregular actions because those effects do not properly belong to [184] the nature of that kind of creatures; for every creature, if regularly made, has particular motions proper to its figure; for natural matter's wisdom makes distinctions by her distinct corporeal motions, giving every particular creature their due portion and proportion according to the nature of their figures and to the rules of her actions, but not to the rules of arts, mathematical compasses, lines, figures, and the like.

---

47. Intelligence: information.

48. A "hundred" was a subdivision of a county.

49. In *Philosophical and Physical Opinions* (1663), Cavendish discusses how nature's parts share information in part 2, chapter 11.

And thus the sun, stars, meteors, air, fire, water, earth, minerals, vegetables, and animals may all have sense and reason, although it does not move in one kind or sort of creatures or in one particular as in another. For the corporeal motions differ not only in kinds and sorts, but also in particulars, as is perceivable by human sense and reason; which is the cause that elements have elemental sense and knowledge, and animals animal sense and knowledge, and so of vegetables, minerals, and the like. Wherefore the sun and stars may have as much sensitive and rational life and knowledge as other creatures, but such is according to the nature of their figures, and not animal, or vegetable, or mineral sense and knowledge. And so leaving them, I rest, Madam,

Your faithful friend and servant.

[185]   **18.**

Madam,

Your author denying that fancy, reason, and animadversion[50] are seated in the brain, and that the brain is figured into this or that conception, "I demand," says he, "in what knot, loop, or interval thereof does this faculty of free fancy and active reason reside?"[51] My answer is that in my opinion, fancy and reason are not made in the brain, as there is a brain, but as there is sensitive and rational matter, which makes not only the brain, but all thoughts, conceptions, imaginations, fancy, understanding, memory, remembrance, and whatsoever motions are in the head or brain; neither does this sensitive and rational matter remain or act in one place of the brain, but in every part thereof; and not only in every part of the brain, but in every part of the body; nay, not only in every part of a man's body, but in every part of nature.

But, madam, I would ask those that say the brain has neither sense, reason, nor self-motion, and therefore no perception, but that all proceeds from an immaterial principle, as an incorporeal spirit distinct from the body which moves and actuates corporeal matter; I would fain ask them, I say, where their immaterial ideas reside, in what part or place of the body? And whether they be little or great? Also I would ask them whether there can be many or but one

---

50. Animadversion: observation.

51. More, *Antidote*, book 1, chapter 11, section 5.

idea of God? If they say many, then there must be several distinct [186] deitical ideas; if but one, where does this idea reside? If they say in the head, then the heart is ignorant of God; if in the heart, then the head is ignorant thereof, and so for all parts of the body; but if they say in every part, then that idea may be disfigured by a lost member; if they say it may dilate and contract, then I say it is not the idea of God, for God can neither contract nor extend; nor can the idea itself dilate and contract, being immaterial; for contraction and dilation belong only to bodies, or material beings.

Wherefore the comparisons between nature and a particular creature, and between God and nature, are improper; much more between God and nature's particular motions and figures, which are various and changeable, although methodical. The same I may ask of the mind of man as I do of the idea in the mind. Also I might ask them what they conceive the natural mind of man to be, whether material or immaterial? If material, their opinion is rational, and so the mind is dividable and composable; if immaterial, then it is a spirit; and if a spirit, it cannot possibly dilate nor contract, having no dimension nor divisibility of parts (although your author proves it by the example of light; but I have expressed my meaning heretofore, that light is divisible[52]); and if it have no dimension, how can it be confined in a material body?

Wherefore when your author says the mind is a substance, it is to my reason very probable; but not when he says it is an immaterial substance, which will never agree with my sense and reason; for it must be either something or nothing, there being no medium between, in nature. But pray mistake me not, Madam, when I say immaterial is nothing; for [187] I mean nothing natural, or so as to be a part of nature; for God forbid I should deny that God is a spiritual immaterial substance or being; neither do I deny that we can have an idea, notion, conception, or thought of the existence of God; for I am of your author's opinion that there is no man under the cope of heaven that does not by the light of nature know and believe there is a God; but that we should have such a perfect idea of God, as of anything else in the world or as of ourselves, as your author says, I cannot in sense and reason conceive to be true or possible.

Neither am I against those spirits which the holy Scripture mentions, as angels and devils, and the divine soul of man; but I say only that no immaterial spirit belongs to nature so as to be a part thereof; for nature is material, or corporeal; and whatsoever is not composed of matter or body belongs not to

---

52. For Cavendish's claim that light is a body, see in this volume, section 1, letters 19–20. In *Philosophical and Physical Opinions* (1663), light is discussed in part 5, chapter 20.

nature; nevertheless, immaterial spirits may be in nature, although not parts of nature. But there can neither be an immaterial nature nor a natural immaterial. Nay, our very thoughts and conceptions of immaterial are material, as made of self-moving matter.

Wherefore to conclude, these opinions in men proceed from a vain-glory, as to have found out something that is not in nature; to which I leave them and their natural immaterial substances, like so many hobgoblins to fright children withal, resting in the meantime, Madam,

Your faithful friend and servant.

[188]   **19.**

Madam,

There are various opinions concerning the seat of the common sense, as your author rehearses them in his treatise *Of the Immortality of the Soul*.[53] But my opinion is that common sense has also a common place; for as there is not any part of the body that has not sense and reason, so sense and reason is in all parts of the body, as it is observable by this, that every part is subject to pain and pleasure, and all parts are moveable, moving, and moved; also appetites are in every part of the body.

As for example, if any part itches, it has an appetite to be scratched, and every part can pattern out several objects, and so several touches; and though the rational part of matter is mixed in all parts of the body, yet it has more liberty to make variety of motions in the head, heart, liver, spleen, stomach, bowels, and the like, than in the other parts of the body; nevertheless, it is in every part, together with the sensitive. But they do not move in every part alike, but differ in each part more or less, as it may be observed; and although every part has some difference of knowledge, yet all have life and knowledge, sense and reason, some more, some less, and the whole body moves according to each part, and so do all the bodily faculties and proprieties, and not according to one single part; the rational soul being in all parts of the body. For if one part of the body should

---

53. More, *Immortality*, book 2, chapter 4. The "common sense" was that part of the mind thought to unify perceptions from the particular sense modalities (vision, hearing, etc.) into a perception of a single entity.

have a dead palsy, [189] it is not that the soul is gone from that part, but that the sensitive and rational matter has altered its motion and figure from animal to some other kind; for certainly, the rational soul, and so life, is in every part, as well in the pores of the skin as in the ventricles of the brain, and as well in the heel as in the head; and every part of the body knows its own office, what it ought to do, from whence follows an agreement of all the parts. And since there is difference of knowledge in every part of one body, well may there be difference between several kinds and sorts, and yet there is knowledge in all; for difference of knowledge is no argument to prove they have no knowledge at all.

Wherefore I am not of the opinion that that which moves the body is as a point, or some such thing in a little kernel or "glandula" of the brain, as an ostrich-egg is hung up to the roof of a chamber;[54] or that it is in the stomach like a single penny in a great purse; neither is it in the midst of the heart, like a lady in a lobster;[55] nor in the blood, like as a minnow or sprat in the sea; nor in the fourth ventricle of the brain, as a lousy[56] soldier in a watch-tower. But you may say, it is like a farthing candle in a great church. I answer, that light will not enlighten the by-chapels of the church, nor the quest-house,[57] nor the belfry; neither does the light move the church, though it enlightens it. Wherefore the soul after this manner does not move the corporeal body, no more than the candle moves the church, or the lady moves the lobster, or the sprat the sea as to make it ebb and flow. [. . .]

[190] [. . .] Nor is there anything which can better prove the mind to be corporeal than that there may be several figures in several parts of the body made at one time, as sight, hearing, tasting, smelling, and touching, and all these in each several organ, as well at one as at several times, either by patterns or not; which figuring without pattern may be done as well by the sensitive motions in the organs as by the rational in the mind, and is called "remembrance." As for example: a man may hear or see without an object; which is, that the sensitive and rational matter repeat such figurative actions, or make others in the sensitive organs or in the mind; and thoughts, memory, imagination, as also passion, are no less corporeal actions than the motion of the hand or heel; neither has the rational matter, being naturally wise, occasion to jumble and knock her parts

---

54. In the Middle Ages, ostrich eggs were occasionally hung in church vaults, possibly as a way to attract people to church to see the rarity. See Creighton Gilbert, "'The Egg Reopened' Again," *The Art Bulletin* 56, no. 2 (1974): 252–58.

55. A small part of a lobster's stomach is thought by some people to look like a lady's head.

56. Lousy: lice-infested.

57. Quest-house: a house where inquests were conducted.

together, by reason every part knows naturally their office, what [191] they ought to do or what they may do.

But I conclude, repeating only what I have said oft before, that all perceptions, thoughts, and the like are the effects, and life and knowledge [are] the nature and essence of self-moving matter. And so I rest, Madam,

Your faithful friend and servant.

[194]   **21.**

Madam,

Your author endeavors very much to prove the existency of a natural immaterial spirit, whom he defines to be an "incorporeal substance, indivisible, that can move itself, can penetrate, contract and dilate itself, and can also move and alter the matter."[58] Whereof, if you will have my opinion, I confess freely to you that in my sense and reason I cannot conceive it to be possible that there is any such thing in nature; for all that is a substance in nature is a body, and what has a body is corporeal; for though there be several degrees of matter, as in purity, rarity, subtlety, activity, yet there is no degree so pure, rare, and subtle, that can go beyond its nature and change from corporeal to incorporeal, except it could change from being something to nothing, which is impossible in nature.

Next, there is no substance in nature that is not divisible; for all that is a body or a bodily substance has extension, and all extension has parts, and what has parts is divisible. As for self-motion, contraction, and dilation, these are actions only of natural matter; for matter by the power of God is self-moving, and all sorts of motions, as contraction, dilation, alteration, penetration, etc., do properly belong to matter; so that natural matter stands in no need to have some immaterial or incorporeal substance to move, rule, guide, and govern her, but she is able enough to do it all herself, by the [195] free gift of the omnipotent God; for why should we trouble ourselves to invent or frame other unconceivable substances, when there is no need for it, but matter can act and move as well

58. More, *Immortality*, book 1, chapter 5, section 1. More's actual definition says incorporeal substance is "indiscerpible" rather than indivisible, but in the sentence before this one, he equates the two.

without them and of itself? Is not God able to give such power to matter as to another incorporeal substance?

But I suppose this opinion of natural immaterial spirits does proceed from chemistry, where the extracts are vulgarly called spirits;[59] and from that degree of matter, which by reason of its purity, subtlety, and activity is not subject to our grosser senses. However, these are not incorporeal, be they never so pure and subtle. And I wonder much that men endeavor to prove immaterial spirits by corporeal arts, when as art is not able to demonstrate nature and her actions; for art is but the effect of nature, and expresses rather the variety than the truth of natural motions; and if art cannot do this, much less will it be able to express what is not in nature or what is beyond nature; as to "trace the visible" (or rather invisible) "footsteps of the divine counsel and providence,"[60] or to demonstrate things supernatural, and which go beyond man's reach and capacity.

But to return to immaterial spirits, that they should rule and govern infinite corporeal matter like so many demigods, by a dilating nod, and a contracting frown, and cause so many kinds and sorts of corporeal figures to arise, being incorporeal themselves, is impossible for me to conceive; for how can an immaterial substance cause a material corporeal substance, which has no motion in itself, to form so many several and various figures and creatures, and make so many alterations, and continue their kinds and sorts by perpetual successions of particulars?

But [196] perchance the immaterial substance gives corporeal matter motion. I answer, my sense and reason cannot understand how it can give motion, unless motion be different, distinct, and separable from it; nay, if it were, yet being no substance or body itself, according to your author's and others' opinion, the question is how can it be transmitted or given away to corporeal matter?

Your author may say that his immaterial and incorporeal spirit of nature, having self-motion, does form matter into several figures. I answer, then that immaterial substance must be transformed and metamorphosed into as many several figures as there are figures in matter; or there must be as many spirits as there are figures; but when the figures change, what does become of the spirits?

Neither can I imagine that an immaterial substance, being without body, can have such a great strength as to grapple with gross, heavy, dull, and dead matter. Certainly, in my opinion, no angel, nor devil, except God empower him, would

---

59. Cavendish is alluding to alchemical processes such as distillation, whereby some physical substance becomes more concentrated or refined.

60. More, *Antidote*, book 2, chapter 2, section 15.

be able to move corporeal matter, were it not self-moving, much less any natural spirit. But God is a spirit, and immoveable; and if created immaterials participate of that nature, as they do of the name, then they must be immoveable also.

Your author, Madam, may make many several degrees of spirits; but certainly not I, nor I think any natural creature else, will be able naturally to conceive them. He may say, perchance, there is such a close conjunction between body and spirit, as I make between rational, sensitive, and inanimate matter. I answer that these degrees are all but one matter, and of one and the same nature as mere matter, different only in degrees of [197] purity, subtlety, and activity, whereas spirit and body are things of contrary natures.

In fine, I cannot conceive how a spirit should fill up a place or space, having no body, nor how it can have the effects of a body, being none itself; for the effects flow from the cause; and as the cause is, so are its effects. And so confessing my ignorance, I can say no more, but rest, Madam,

Your faithful friend and servant.

22.

Madam,

Your author having assigned indivisibility to the soul or spirit that moves and actuates matter, I desire to know how one indivisible spirit can be in so many dividable parts? For there being infinite parts in nature, they must either have one infinite spirit to move them, which must be dilated infinitely, or this spirit must move severally[61] in every part of nature. If the first, then I cannot conceive but all motion must be uniform, or after one and the same manner; nay, I cannot understand how there can be any dilation and contraction, or rather any motion of the same spirit, by reason if it dilate, then (being equally spread out in all the parts of matter) it must dilate beyond matter; and if it contract, it must leave some parts of matter void and without [198] motion. But if the spirit moves every part severally, then he is divisible; neither can I think that there are so many spirits as there are parts in nature; for your author says there is but one spirit of nature. I will give an easy and plain example: when a worm is cut into two or three parts, we see there is sensitive life and motion in every part, for

---

61. Severally: separately.

every part will strive and endeavor to meet and join again to make up the whole body; now if there were but one indivisible life, spirit, and motion, I would fain know how these severed parts could move all by one spirit.

Wherefore, matter, in my opinion, has self-motion in itself, which is the only soul and life of nature, and is dividable as well as composable, and full of variety of action; for it is as easy for several parts to act in separation as in composition, and as easy in composition as in separation. Neither is every part bound to one kind or sort of motions; for we see in exterior local motions that one man can put his body into several shapes and postures, much more can nature.

But is it not strange, Madam, that a man accounts it absurd, ridiculous, and a prejudice to God's omnipotency to attribute self-motion to matter or a material creature, when it is not absurd, ridiculous, or any prejudice to God to attribute it to an immaterial creature? What reason of absurdity lies herein? Surely I can conceive none, except it be absurd and ridiculous to make that which no man can know or conceive what it is, *viz.*, an immaterial natural spirit (which is as much as to say, a natural nothing) to have motion, and not only motion, but self-motion; nay, not only self-motion, but to move, actuate, rule, govern, and guide matter, [199] or corporeal nature, and to be the cause of all the most curious varieties and effects in nature. Was not God able to give self-motion as well[62] to a material as to an immaterial creature, and endow matter with a self-moving power?

I do not say, Madam, that matter has motion of itself so that it is the prime cause and principle of its own self-motion; for that were to make matter a god, which I am far from believing; but my opinion is that the self-motion of matter proceeds from God, as well as the self-motion of an immaterial spirit; and that I am of this opinion, the last chapter of my book of philosophy will inform you, where I treat of the deitical center as the fountain from whence all things do flow, and which is the supreme cause, author, ruler, and governor of all.[63]

Perhaps you will say it is because I make matter eternal. 'Tis true, Madam, I do so. But I think eternity does not take off the dependence upon God, for God may nevertheless be above matter, as I have told you before. You may ask me how that can be? I say, as well as anything else that God can do beyond our understanding. For I do but tell you my opinion, that I think it most probable to be so, but I can give you no mathematical demonstrations for it. Only this I am

---

62. Well: easily.

63. In the last chapter of *Philosophical and Physical Opinions* (1663), Cavendish says God is to infinite matter as the center of an infinitely large circle is to the circle itself.

sure of, that it is not impossible for the omnipotent God; and he that questions the truth of it may question God's omnipotency.

Truly, Madam, I wonder how man can say God is omnipotent and can do beyond our understanding, and yet deny all that he is not able to comprehend with his reason. However, as I said, it is my opinion that matter is self-moving by the power of God. Neither can animadversion and perception, as also the [200] variety of figures, prove that there must be another eternal agent or power to work all this in matter; but it proves rather the contrary; for were there no self-motion in matter, there would be no perception, nor no variety of creatures in their figures, shapes, natures, qualities, faculties, proprieties, as also in their productions, creations or generations, transformations, compositions, dissolutions, and the like, as growth, maturity, decay, etc., and for animals, were not corporeal matter self-moving, dividable, and composable, there could not be such variety of passions, complexions, humors, features, statures, appetites, diseases, infirmities, youth, age, etc. Neither would they have any nourishing food, healing salves, sovereign[64] medicines, reviving cordials, or deadly poisons.

In short, there is so much variety in nature proceeding from the self-motion of matter as not possible to be numbered, nor thoroughly known by any creature. Wherefore I should labor in vain if I endeavored to express any more thereof; and this is the cause that I break off here, and only subscribe myself, Madam,

Your faithful friend and servant.

[204] ## 24.

Madam,

Having given you my opinion both of the substance and perception of light in my last letter, I perceive your desire to know how shadows are made. Truly, Madam, to my sense and reason, it appears most probable that shadows are made by the way of patterning. As for example, when a man's or tree's or any other the like creature's shadow is made upon the ground or the wall or the like, those bodies as the ground or wall do, in my opinion, pattern out the interposing body that is between the light and them. And the reason that the shadow is

---

64. Sovereign: in a medical context, highly potent or effective.

longer or shorter, or bigger or less, is according as the light is nearer or further off; for when the light is perpendicular, the interposing body cannot obscure the light, because the light surrounding the interposing body by its brightness rather obscures the body, than the body the light; for the numerous and splendorous [205] patterns of light taken from the body of the sun do quite involve the interposing body.

Next, you desire to know whether the light we see in the moon, be the moon's own natural light, or a borrowed light from the sun. I answer that in my opinion, it is a borrowed light; to wit, that the moon does pattern out the light of the sun. And the proof of it is that when the sun is in an eclipse, we do plainly perceive that so much of the sun is darkened as the moon covers; for though those parts of the moon that are next [to] the sun may, for anything we know, pattern out the light of the sun, yet the moon is dark on that side which is from the sun. I will not say but that part of the moon which is towards the earth may pattern out the earth, or the shadow of the earth, which may make the moon appear more dark and sullen. But when the moon is in an eclipse, then it is plainly perceived that the moon patterns out the earth, or the shadow of the earth. Besides, those parts of the moon that are farthest from the sun are dark, as we may observe when as the moon is in the wane, and enlightened when the sun is nearer. But I will leave this argument to observing astrologers, and rest, Madam,

Your faithful friend and servant.

[206]   **25.**

Madam,

If according to your author's opinion, "In every particular world, such as man is especially, his own soul" (which is a spirit) "be the peculiar and most perfective architect of the fabric of his body, as the soul of the world is of it,"[65] then I cannot conceive in my reason how the separation is made in death; for I see that all animals, and so mankind, have a natural desire to live, and that life and soul are unwilling to part. And if the power lies in the soul, why does she not continue with the body, and animate, move, and actuate it, as she did before, or order the

---

65. More, *Immortality*, book 2, chapter 10, section 2.

matter so as not to dissolve? But if the dissolution lies in the body, then the body has self-motion.

Yet it is most probable, if the soul be the architect of the body, it must also be the dissolver of it; and if there come not another soul into the parts of matter, the body must either be annihilated or lie unmoved as long as the world lasts, which is improbable; for surely all the bodies of men or other animals are employed by nature to some use or other. However, it is requisite that the soul must stay so long in the body, until it be turned into dust and ashes; otherwise, the body having no self-motion, would remain as it was when the soul left it, that is, entire and undissolved. As for example, when a man dies, if there be no motion in his body, and the soul, which was the mover, be [207] gone, it cannot possibly corrupt;[66] for certainly, that we call "corruption" is made by motion, and the body requires as much motion to be dissolved or divided as it does to be framed or composed. Wherefore a dead body would remain in the same state continually if it had no self-motion in it. And if another soul should enter into the body and work it to another figure, then certainly there must be much more souls than bodies, because bodies are subject to change into several forms; but if the animal spirits, which are left in the body after the soul is gone, are able to dissolve it without the help of the soul, then it is probable they could have framed it without the help of the soul; and so they being material, it must be granted that matter is self-moving.

But if corporeal matter have corporeal self-motion, a self-moving immaterial spirit, by reason of their different natures, would make great obstruction, and so a general confusion; for the corporeal and incorporeal motions would hinder and oppose each other, their natures being quite different; and though they might subsist together without disturbance of each other, yet it is not probable they should act together, and that in such a conjunction as if they were one united body; for it is, in my opinion, more probable that one material should act upon another material, or one immaterial upon another immaterial, than that an immaterial should act upon a material or corporeal.

Thus the consideration or contemplation of immaterial natural spirits puts me always into doubts and raises so many contradictions in my sense and reason, as I know not, nor am not [208] able to reconcile them. However, though I am doubtful of them, yet I can assure yourself that I continue, Madam,

> Your faithful friend and servant.

---

66. Corrupt: decompose.

27.

Madam,

Your author in the continuation of his discourse concerning the immaterial soul of man, demonstrating that her seat is not bound up in a certain place of the body, but that she pervades all the body and [212] every part thereof, takes, amongst the rest, an argument from passions and sympathies. "Moreover," says he, "passions and sympathies, in my judgment, are more easily to be resolved into this hypothesis of the soul's pervading the whole body, than in restraining its essential presence to one part thereof.—But it is evident that they arise in us against both our will and appetite. For who would bear the tortures of fears and jealousies, if he could avoid it?"[67]

Concerning passions, Madam, I have given my opinion at large in my book of philosophy,[68] and am of your author's mind that passions are made in the heart, but not by an immaterial spirit, but by the rational soul, which is material; and there is no doubt but that many passions, as fear, jealousy, etc., arise against our will and appetite; for so many foreign nations invade any kingdom without the will or desire of the inhabitants, and yet they are corporeal men. The same may be said of passions; and several parts of matter may invade each other, whereof one may be afraid of the other, yet all this is but according as corporeal matter moves, either generally or particularly. Generally, that is, when many parts of matter unite or join together, having the like appetites, wills, designs; and we may observe, that there are general agreements amongst several parts in plagues as well as wars, which plagues are not only amongst men, but amongst beasts; and sometimes but in one sort of animals, as a general rot amongst sheep, a general mange amongst dogs, a general farcy[69] amongst horses, a general plague amongst men; all which could not be without a general infection, one part infecting another, or rather one part imitating the motions of the other [213] that is next adjoining to it; for such infections come by the near adhesion of parts, as is observable, which immaterial and individable natural spirits could not effect; that is, to make such a general infection in so many several parts of so

---

67. More, *Immortality*, book 2, chapter 10, section 6.

68. I.e., her *Philosophical and Physical Opinions*. In the 1663 edition, Cavendish discusses passions in part 6, chapter 7, where she says that the hypothesis that passions arise in the heart rather than the mind explains how there can be conflicts between emotion and judgment; how one can feel an emotion without knowing why; and how one can become mentally unbalanced yet still feel the same passions as before.

69. Farcy: a bacterial disease primarily affecting horses.

many several creatures, to the creatures' dissolution. Also there will be several invasions at one time, as plague and war, amongst neighboring and adjoining creatures or parts.

But this is to be observed, that the sensitive corporeal motions make all diseases, and not the rational, although the rational are many times the occasion that the sensitive do move into such or such a disease; for all those that are sick by conceit,[70] their sicknesses are caused by the rational corporeal motions. But being loathe to make tedious repetition hereof, having discoursed of diseases and passions in my mentioned book of philosophy,[71] I refer you thither, and rest, Madam,

Your faithful friend and servant.

28.

Madam,

Concerning dimness of sight, which your author will have to "proceed from the deficiency of the animal spirits,"[72] my meaning in short is that when sight is dim, though the sensitive organs are perfect, [214] this dimness is caused by the alteration only of the sensitive motions in the organs not moving to the nature of sight. And so is made deafness, dumbness, lameness, and the like, as also weariness; for the relaxation of strength in several parts is only an alteration of such sorts of motions which make the nerves strong; and if a man be more dull at one time than at another, it is that there are not so many changes of motions, nor so quick motions at that time as at another; for nature may use more or less force as she pleases. Also she can and does often use opposite actions, and often sympathetical and agreeable actions, as she pleases; for nature, having a free power to move, may move as she will; but being wise, she moves as she thinks best, either in her separating or uniting motions, for continuance as well as for variety.

But if, according to your author, the immaterial soul should determinate matter in motion, it would, in my opinion, make a confusion; for the motions

---

70. Conceit: imagination.

71. On the passions, see *Philosophical and Physical Opinions* (1663), part 6, chapters 7–9. Part 7 is devoted to discussions of disease.

72. More, *Immortality*, book 2, chapter 8, section 9.

of the matter would often oppose and cross the motions of the immaterial soul, and so they would disagree, as a king and his subjects (except God had given the soul an absolute power of command, and restrained matter to an irresistible and necessitated obedience; which, in my opinion, is not probable). By which disagreement, nature, and all that is in nature, would have been quite ruined at this time; for no kinds, sorts, or particulars would keep any distinction, if matter did not govern itself, and if all the parts did not know their own affairs, abilities, offices, and functions. Besides, it would, to my thinking, take up a great deal of time to receive commands in every several action, at least so much that, [215] for example, a man could not have so many several thoughts in so short a time as he has.

But concerning the animal spirits, which your author calls the instruments, organs, and engines of the incorporeal soul; I would fain know whether they have no motion but what comes from the soul, or whether they have their own motion of themselves? If the first, then the soul must, in my opinion, be like a deity, and have a divine power to give and impart motion; if the second, then the spirits being material, it follows that matter has motion of itself, or is self-moving. But if the immaterial natural soul can transfer her gifts upon corporeal matter, then it must give numerous sorts of motions, with all their degrees; as also the faculty of figuring, or moving figuratively in all corporeal matter. Which power, to my judgment, is too much for a creature to give.

If you say, the immaterial soul has this power from God; I answer, matter may have the same, and I cannot imagine why God should make an immaterial spirit to be the proxy or vice-gerent[73] of his power, or the "quarter-master general of the divine providence," as your author is pleased to style it,[74] when he is able to effect it without any under-officers, and in a more easy and compendious[75] way, as to impart immediately such self-moving power to natural matter, which man attributes to an incorporeal spirit.

But to conclude, if the animal spirits be the instruments of the incorporeal soul, then the spirits of wine are more powerful than the animal spirits, nay, than the immaterial soul herself; for they can put them and all their actions quite [216] out of order; the same may be done by other material things, vegetables,

---

73. Vice-gerent: someone appointed by a ruler to act as the ruler's representative.

74. More, *Immortality*, book 3, chapter 13, section 10.

75. Compendious: economical.

minerals, and the like. And so leaving this discourse to your better consideration, I take my leave for this time, and rest, Madam,

Your faithful and affectionate friend and servant.

29.

Madam,

Touching the state or condition of the supernatural and divine soul both in and after this life, I must crave your excuse that I can give no account of it; for I dare affirm nothing; not only that I am no professed divine and think it unfit to take anything upon me that belongs not to me, but also that I am unwilling to mingle divinity and natural philosophy together, to the great disadvantage and prejudice of either; for if each one did contain himself within the circle of his own profession, and nobody did pretend to be a divine philosopher, many absurdities, confusions, contentions, and the like would be avoided, which now disturb both church and schools, and will in time cause their utter ruin and destruction. For what is supernatural cannot be naturally known by any natural creature; neither can any supernatural creature but the infinite and eternal God know [217] thoroughly everything that is in nature, she being the infinite servant of the infinite God, whom no finite creature of what degree soever, whether natural or supernatural, can conceive; for if no angel nor devil can know our thoughts, much less will they know infinite nature; nay, one finite supernatural creature cannot, in my opinion, know perfectly another supernatural creature, but God alone, who is all-knowing.

And therefore all what is said of supernatural spirits, I believe, so far as the Scripture makes mention of them; further I dare not presume to go; the like of the supernatural or divine soul. For all that I have writ hitherto to you of the soul concerns the natural soul of man, which is material, and not the supernatural or divine soul; neither do I contradict anything concerning this divine soul, but am only against those opinions which make the natural soul of man an immaterial natural spirit, and confound supernatural creatures with natural, believing those spirits to be as well natural creatures and parts of nature as material and corporeal beings are; when as there is great difference between them, and nothing in nature to be found but what is corporeal.

Upon this account I take all their relations of demons, of the genii, and of the souls after the departure from human bodies, their vehicles, shapes, habitations, converses,[76] conferences, entertainments, exercises, pleasures, pastimes, governments, orders, laws, magistrates, officers, executioners, punishments, and the like, rather for poetical fancies, than truth and reason, whether they concern the divine or natural soul. For as for the divine soul, the Scripture [218] makes no other mention of it, but that immediately after her departure out of this natural life, she goes either to heaven or hell, either to enjoy reward or to suffer punishment, according to man's actions in this life.

But as for the natural soul, she, being material, has no need of any vehicles, neither is natural death anything else but an alteration of the rational and sensitive motions, which from the dissolution of the one figure go to the formation or production of another. Thus the natural soul is not like a traveller, going out of one body into another, neither is air her lodging;[77] for certainly, if the natural human soul should travel through the airy regions, she would at last grow weary, it being so great a journey, except she did meet with the soul of a horse and so ease herself with riding on horseback.

Neither can I believe souls or demons in the air have any commonwealth, magistrates, officers, and executioners in their airy kingdom; for wheresoever are governments, magistrates, and executioners, there are also offenses, and where there is power to offend as well as to obey, there may and will be sometimes rebellions and civil wars; for there being different sorts of spirits, it is impossible they should all so well agree, especially the good and evil genii, which certainly will fight more valiantly than Hector and Achilles; nay, the spirits of one sort would have more civil wars than ever the Romans had; and if the soul of Caesar and Pompey should meet, there would be a cruel fight between those two heroical souls; the like between Augustus's and Antonius's soul.

But, Madam, all these, as I said, I take for fancies proceeding from the religion of the gentiles,[78] not fit for Christians [219] to embrace for any truth; for if we should, we might at last, by avoiding to be atheists, become pagans, and so leap out of the frying-pan into the fire, as turning from divine faith to poetical fancy; and if Ovid should revive again, he would, perhaps, be the chief head or pillar of the church.

---

76. Converses: conversations.

77. On the idea that the soul lodges in the air, see the Introduction (p. xxxv–xxxvi).

78. Gentiles: pagans, heathens.

By this you may plainly see, Madam, that I am no Platonic;[79] for this opinion is dangerous, especially for married women, by reason the conversation of the souls may be a great temptation, and a means to bring platonic lovers to a nearer acquaintance, not allowable by the laws of marriage, although by the sympathy of the souls. But I conclude, and desire you not to interpret amiss this my discourse, as if I had been too invective against poetical fancies; for that I am a great lover of them, my poetical works will witness;[80] only I think it not fit to bring fancies into religion. Wherefore what I have writ now to you, is rather to express my zeal for God and his true worship, than to prejudice anybody; and if you be of that same opinion, as above mentioned, I wish my letter may convert you, and so I should not account my labor lost, but judge myself happy that any good could proceed to the advancement of your soul, from, Madam,

Your faithful friend and servant.

[220]   **30.**

Madam,

I sent you word in my last, I would not meddle with writing anything of the divine soul of man, by reason it belongs to faith and religion and not to natural philosophy; but since you desire my opinion concerning the immortality of the divine soul, I cannot but answer you plainly that first I did wonder much you made question of that whose truth, in my opinion, is so clear as hardly any rational man will make a doubt of it; for I think there is almost no Christian in the world but believes the immortality of the soul, no not Christians only, but Mahometans[81] and Jews. But I left[82] to wonder at you when I saw wise and learned men, and great divines, take so much pains as to write whole volumes, and bring so many arguments to prove the immortality of the soul; for this was a greater miracle to me than if nature had showed me some of her secret and hidden effects, or if I had seen an immaterial spirit.

---

79. By "Platonic" here, Cavendish means an adherent of the belief that souls are distinct from bodies. In the next clause, she refers to "platonic lovers," meaning those whose love is spiritual rather than physical.

80. I.e., her 1653 book *Poems, and Fancies*.

81. Mahometans: Muslims.

82. Left: ceased.

Certainly, Madam, it seems as strange to me to prove the immortality of the soul as to convert atheists; for it [is] impossible, almost, that any atheist should be found in the world. For what man would be so senseless as to deny a God? Wherefore to prove either a God or the immortality of the soul is to make a man doubt of either. For as physicians and surgeons apply strengthening medicines only to those parts of the body which they suppose [221] the weakest, so it is with proofs and arguments, those being for the most part used in such subjects the truth of which is most questionable.

But in things divine, disputes do rather weaken faith than prove truth, and breed several strange opinions; for man being naturally ambitious, and endeavoring to excel each other, will not content himself with what God has been pleased to reveal in his holy word, but invents and adds something of his own; and hence arise so many monstrous expressions and opinions that a simple man is puzzled, not knowing which to adhere to, which is the cause of so many schisms, sects, and divisions in religion. Hence it comes also that some pretend to know the very nature and essence of God, his divine counsels, all his actions, designs, rules, decrees, power, attributes, nay, his motions, affections, and passions, as if the omnipotent infinite God were of a human shape; so that there are already more divisions than religions, which disturb the peace and quiet both of mind and body; when as the ground of our belief consists in some few and short articles, which clearly explained, and the moral part of divinity well pressed upon the people, would do more good than unnecessary and tedious disputes, which rather confound religion than advance it. But if man had a mind to show learning and exercise his wit, certainly there are other subjects wherein he can do it with more profit and less danger than by proving Christian religion with natural philosophy, which is the way to destroy them both.

I could wish, Madam, that everyone would but observe the command of Christ, and give to God [222] what is God's, and to Caesar what is Caesar's, and so distinguish what belongs to the actions of nature, what to the actions of religion; for it appears to my reason that God has given nature, his eternal servant, a peculiar freedom of working and acting as a self-moving power from eternity; but when the omnipotent God acts, he acts supernaturally, as beyond nature; of which divine actions none but the holy church, as one united body, mind, and soul, should discourse, and declare the truth of them, according to the revelation made by God in his holy word, to her flock the laity, not suffering any one single person, of what profession or degree[83] soever, indifferently to

---

83. Degree: social class.

comment, interpret, explain, and declare the meaning or sense of the Scripture after his own fancy.

And as for nature's actions, let those whom nature has endued with such a proportion of reason as is able to search into the hidden causes of natural effects contemplate freely, without any restraint or confinement; for nature acts freely, and so may natural creatures, and amongst the rest man, in things which are purely natural; but as for things supernatural, man cannot act freely, by reason they are beyond his sphere of conception and understanding, so as he is forced to set aside reason and only to work by faith.

And thus, Madam, you see the cause why I cannot give you a full description of the divine soul of man, as I mentioned already in my last, but that I do only send you my opinion of the natural soul, which I call the rational soul; not that I dare say the supernatural soul is without natural reason, but natural reason is not the divine soul; neither can natural reason, without faith, advance the divine soul [223] to heaven, or beget a pious zeal, without divine and supernatural grace. Wherefore reason, or the rational soul, is only the soul of nature, which being material, is dividable, and so becomes numerous in particular natural creatures; like as the sensitive life, being also material and dividable, becomes numerous, as being in every creature and in every part of every creature; for as there is life in every creature, so there is also a soul in every creature; nay, not only in every creature but in every particle of every creature, by reason every creature is made of rational and sensitive matter; and as all creatures or parts of nature are but one infinite body of nature, so all their particular souls and lives make but one infinite soul and life of nature; and this natural soul has only natural actions, not supernatural; nor has the supernatural soul natural actions; for although they subsist both together in one body, yet each works without disturbance to the other; and both are immortal; for of the supernatural soul there is no question, and of the natural soul, I have said before that nothing is perishable or subject to annihilation in nature, and so no death, but what is called by the name of death is only an alteration of the corporeal natural motions of such a figure to another figure; and therefore as it is impossible that one part of matter should perish in nature, so is it impossible that the natural or rational soul can perish, being material. The natural human soul may alter so as not to move in an animal way, or not to have animal motions, but this does not prove her destruction or anni-hilation, but only a change of the animal figure and its motions, all remaining still in nature.

Thus my faith of the [224] divine, and my opinion of the natural soul, is, that they are both immortal; as for the immediate actions of the divine soul, I leave you to the church, which are the ministers of God and the faithful dispensers of the sacred mysteries of the Gospel, the true expounders of the word of God, reformers of men's lives, and tutors of the ignorant, to whom I submit myself in all that belongs to the salvation of my soul and the regulating of the actions of my life, to the honor and glory of God.

And I hope they will not take any offense at the maintaining and publishing [of] my opinions concerning nature and natural effects, for they are as harmless and as little prejudicial to them as my designs; for my only and chief design is and ever has been to understand nature rightly, obey the church exactly, believe undoubtedly, pray zealously, live virtuously, and wish earnestly that both church and schools may increase and flourish in the sacred knowledge of the true word of God, and that each one may live peaceable and happily in this world, die quietly, and rise blessedly and gloriously to everlasting life and happiness. Which happiness I pray God also to confer upon your Ladyship. Till then, I rest, Madam,

Your faithful and constant friend, to serve you.

[225] **31.**

Madam,

I will leave the controversy of free will and necessity, which your author is discoursing of,[84] to divines to decide it, only I say this, that nature has a natural free will and power of self-moving, and is not necessitated; but yet that this free will proceeds from God, who has given her both will and power to act freely.

But as for the question, whether there be nothing in the universe, but mere body?[85] I answer, my opinion is not that there is nothing in the world but mere body; but that nature is purely material or corporeal, and that there is no part of nature, or natural creature, which is not matter or body, or made of matter; also, that there is not anything else mixed with body, as a co-partner in natural actions, which is distinct from body or matter; nevertheless, there may be supernatural spiritual beings or substances in nature without any hindrance to matter

---

84. More, *Immortality*, book 2, chapter 3. (Cavendish misidentifies this as book 1, an unsurprising mistake since the 1662 book itself has a typographical error in the header.)

85. More, *Immortality*, book 2, chapter 2.

or corporeal nature. The same I may say of the natural material, and the divine and supernatural soul; for though the divine soul is in a natural body, and both their powers and actions be different, yet they cause no ruin or disturbance to each other, but do in many cases agree with each other, without encroachment upon each other's powers or actions; for God, as he is the God of all things, so the God of order. Wherefore it is not probable that created immaterial or incorporeal beings [226] should order corporeal nature, no more than corporeal nature orders immaterial or incorporeal creatures.

Neither can, in my opinion, incorporeal creatures be clearly conceived by corporeals, although they may really exist and subsist in nature; only, as I said before, it is well to be considered that there is difference between being in nature and being a part of nature; for bodiless things, and so spiritual substances, although they may exist in nature, yet they are not natural nor parts of nature, but supernatural, nature being merely corporeal, and matter the ground of nature; and all that is not built upon this material ground is nothing in nature.

But you will say, the divine soul is a part of man, and man a part of nature, wherefore the divine soul must needs be a part of nature. I answer, not. For the divine soul is not a part of nature, but supernatural; as a supernatural gift from God only to man, and to no other creature. And although in this respect it must be called a part of man, yet it is no natural or material part of man; neither does this supernatural gift disturb nature or natural matter, or natural matter this supernatural gift. And so leaving them both, I rest, Madam,

Your faithful friend and servant.

SECTION 3

[234]   1.

Madam,

Having discharged my duty thus far, that in obedience to your commands, I have given you my answers to the opinions of three of those famous and learned authors you sent me, *viz.*, Hobbes, Descartes, and More, and explained my own opinions by examining theirs; my only task shall be now to proceed in the same manner with that famous philosopher and chemist, van Helmont.[1] But him I find more difficult to be understood than any of the forementioned, not only by reason of the art of chemistry, which I confess myself not versed in, but especially that he has such strange terms and unusual expressions as may puzzle anybody to apprehend the sense and meaning of them. Wherefore, if you receive not that full satisfaction you expect from me, in examining his opinions and [235] arguments, I beg your pardon beforehand, and desire you to remember that I sent you word in the beginning, I did undertake this work more out of desire to [make] clear my own opinions than a quarrelsome humor to contradict others; which if I do but obtain, I have my aim.

And so to the business. When as your author discourses of the causes and beginnings of natural things, he is pleased to say that "souls and lives, as they know no degrees, so they know no parts,"[2] which opinion is very different from mine. For although I confess that there is but one kind of life and one kind of soul in nature, which is the sensitive life and the rational soul, both consisting not only of matter but of one kind of matter, to wit, animate; nevertheless they are of different degrees, the matter of the rational soul being more agile, subtle, and active, than the matter of the sensitive life; which is the reason that the rational can act in its own substance or degree of matter, and make figures in itself

---

1. Cavendish is discussing Jan Baptist van Helmont's book *Oriatrike, or, Physick Refined* (1662). Her quotations are not always verbatim, but I have only indicated this when Cavendish's alterations affect the meaning. On van Helmont, see the Introduction (pp. xxxvii–xli).

2. Van Helmont, *Oriatrike*, chapter 4, "Of the Causes and Beginnings of Natural Things."

129

and its own parts; when as the sensitive, being of somewhat a grosser degree than the rational, and not so subtle and active, is confined to work with and upon the inanimate matter.

But mistake me not, Madam, for I make only a difference of the degrees of subtlety, activity, agility, purity, between rational and sensitive matter; but as for the rational matter itself, it has no degrees of purity, subtlety, and activity in its own nature or parts, but is always one and the same in its substance in all creatures, and so is the sensitive.

You will ask me, how comes then the difference of so many parts and creatures in nature, if there be no degrees of purity, activity, and subtlety in the substance of the rational and in the [236] substance of the sensitive matter? As for example: if there were no such degrees of the parts of rational matter amongst themselves, as also of the parts of the sensitive, there would be no difference between animals, vegetables, minerals, and elements, but all creatures would be alike without distinction, and have the same manner of sense and reason, life and knowledge.

I answer that although each sort or degree of animate matter, rational as well as sensitive, has in itself or its own substance no degrees of purity, rarity, and subtlety, but is one and the same in its nature or essence, nevertheless each has degrees of quantity, or parts, which degrees of quantity do make the only difference between the several creatures or parts of nature, as well in their general as particular kinds; for both the rational and sensitive matter being corporeal, and so dividable into parts, some creatures do partake more, some less of them, which makes them to have more or less, and so different sense and reason, each according to the nature of its kind. Nay, this difference of the degrees of quantity or parts in the substance of the rational and sensitive matter makes also the difference between particulars in every sort of creatures, as for example, between several particular men. But as I said, the nature or essence of the sensitive and rational matter is the same in all; for the difference consists not in the nature of matter, but only in the degrees of quantity and parts of matter, and in the various and different actions or motions of this same matter. And thus matter being dividable, there are numerous lives and souls in nature, according to the variousness of her several parts and creatures.

Next your author, [237] mentioning the "causes and principles of natural bodies," assigns two first or chief beginnings and corporeal causes of every creature, to wit, the element of water and the "ferment or leaven"; which ferment he calls a "formal creature being"; neither a substance nor an accident, but a neutral thing.[3] Truly, Madam, my reason is not able to conceive this neutral being; for it must either be something or nothing in nature. And if he makes it anything between both, it is a strange monster, and will produce monstrous effects. And as for water, if he does make it a principle of natural things, I see no reason why he excludes the rest of the elements.[4] But, in my opinion, water and the rest of the elements are but effects of nature, as other creatures are, and so cannot be prime causes. The like the ferment, which, to my sense and reason, is nothing else but a natural effect or natural matter.

Concerning his opinion that causes and beginnings are all one, or that there is but little difference between them,[5] I do readily subscribe unto it; but when he speaks of those "things which are produced without life,"[6] my reason cannot find out what or where they should be; for certainly, in nature they are not, nature being life and soul herself, and all her parts being enlivened and soulified, so that there can be no generation or natural production without life.

Neither is my sense and reason capable to understand his meaning when he says that the "seeds of things, and the spirits, as the dispensers thereof, are divided from the material cause."[7] For I do see no difference between the seed and the material cause, but they are all one thing, it being undeniable that the seed is the matter of that which is produced. But your author was pleased [238] to say heretofore that there are but two beginnings or causes of natural things, and now he makes so many more; for, says he, "of efficient and seminal causes, some are efficiently effecting, and others effectively effecting."[8] Which nice[9] distinctions, in my opinion, do but make a confusion in natural knowledge, setting a man's brain on the rack; for who is to conceive all those chimeras and fancies of the "archeus," "ferment," various "ideas," "blas," "gas," and many more,[10] which are neither something nor nothing in nature, but

---

3. Van Helmont, *Oriatrike*, chapter 4.

4. I.e., earth, air, and fire.

5. Van Helmont, *Oriatrike*, chapter 4.

6. Van Helmont, *Oriatrike*, chapter 4.

7. Van Helmont, *Oriatrike*, chapter 4.

8. Van Helmont, *Oriatrike*, chapter 4.

9. Nice: slight.

10. For discussion of these terms, see the Introduction (pp. xxxix–xl).

between both, except a man have the same fancies, visions, and dreams your author had?

Nature is easy to be understood, and without any difficulty, so as we stand in no need to frame so many strange names, able to fright anybody. Neither do natural bodies know many prime causes and beginnings, but there is but one only chief and prime cause from which all effects and varieties proceed, which cause is corporeal nature, or natural self-moving matter, which forms and produces all natural things; and all the variety and difference of natural creatures arises from her various actions, which are the various motions in nature; some whereof are regular, some irregular. I mean irregular as to particular creatures, not as to nature herself, for nature cannot be disturbed or discomposed, or else all would run into confusion. Wherefore irregularities do only concern particular creatures, not infinite nature; and the irregularities of some parts may cause the irregularities of other parts, as the regularities of some parts do cause the regularities of others. And thus according as regularities and irregularities have power, they cause either peace or war, sickness or health, [239] delight and pleasure, or grief and pain, life or death, to particular creatures or parts of nature; but all these various actions are but various effects, and not prime causes; which is well to be observed, lest we confound causes with effects. And so leaving this discourse for the present, I rest, Madam,

Your faithful friend and servant.

2.

Madam,

It is no wonder your author has so many odd and strange opinions in philosophy, since they do not only proceed from strange visions, apparitions, and dreams, but are built upon so strange grounds and principles as "ideas," "archeus," "gas," "blas," "ferment," and the like, the names of which sound so harsh and terrifying as they might put anybody easily into fright, like so many hobgoblins or immaterial spirits; but the best is, they can do no great harm, except it be to trouble the brains of them that love to maintain those opinions; for though they are thought to be powerful beings, yet not being corporeal substances, I cannot imagine wherein their power should consist; for nothing can do nothing.

But to mention each apart; first his "archeus" he calls "the spirit of life, a vital gas or light."[11] [240] [. . .] In the next place, "gas" and "blas" are to your author also true principles of natural things, for "gas is the vapor into which water is dissolved by cold, but yet it is a far more fine and subtle thing than vapor,"[12] which he demonstrates by the art of chemistry. This "gas" in another place he calls a "wild spirit, or breath, unknown hitherto, which can neither be constrained by vessels, not reduced into a visible body; in some things it is nothing but water; as for example in salt, in fruits, and the like."[13] But "blas proceeds from the local and alterative motion of the stars, and is the general beginning of motion, producing heat and cold, and that especially with the changing of the winds."[14] There is also "blas in all sublunary things";[15] witness amulets or preserving pomanders,[16] whereby they do constrain objects to obey them [. . .] [241] [. . .] The "ferment" he describes to be "a true principle or original beginning of things, to wit, a formal created being which is neither a substance, nor an accident, but a neutral being [. . .]."[17] Lastly, his "ideas" are "certain formal seminal lights, mutually piercing each other without the adultery of union; for," says he, "although at first, that which is imagined is nothing but a mere being of reason, yet it does not remain such . . . and therefore that idea is made a spiritual or seminal powerful being, to perform things of great moment."[18] [. . .]

Thus, Madam, I have made a rehearsal of your author's strange and hitherto unknown principles (as his confession is) of natural things, which to my sense and reason are so obscure, intricate, and perplex as is almost impossible exactly to conceive them; [242] when as principles ought to be easy, plain, and without any difficulty to be understood. Wherefore what with his spirits, mere-beings, non-beings, and neutral-beings, he troubles nature and puzzles the brains of his readers so that, I think, if all men were of his opinion or did follow the way of his philosophy, nature would desire God she might be annihilated. Only, of all other, she does not fear his non-beings, for they are the weakest of all and can do her the least hurt, as not being able to obstruct real and corporeal actions

---

11. Van Helmont, *Oriatrike*, chapter 18, "The Fiction of Elementary Complexions and Mixtures."

12. Van Helmont, *Oriatrike*, chapter 13, "Of the Gas of the Water."

13. Van Helmont, *Oriatrike*, chapter 18, "The Fiction of Elementary Complexions and Mixtures."

14. Van Helmont, *Oriatrike*, chapter 14, "Of the Blas of Meteors."

15. Van Helmont, *Oriatrike*, chapter 42, "An Unknown Action of Government."

16. Pomander: a ball composed of aromatic herbs, spices, or other fragrant substances, carried in a pocket or worn around the neck and thought to ward off disease.

17. Van Helmont, *Oriatrike*, chapter 4, "The Causes and Beginnings of Natural Things."

18. Van Helmont, *Oriatrike*, chapter 69, "Of the Ideas of Diseases."

of nature; for nature is a corporeal substance, and without a substance motion cannot be, and without motion opposition cannot be made, nor any action in nature, whether prints, seals, stamps, productions, generations, thoughts, conceptions, imaginations, passions, appetites, or the like. And if motions cannot be without substance, then all creatures, their properties, faculties, natures, etc. being made by corporeal motions, cannot be non-beings, no nor anything else that is in nature; for non-beings are not in the number of natural things, nature containing nothing within her but what is substantially, really, and corporeally existent [. . .].

[243] [. . .] But to return to "ideas." I had almost forgot to tell you, Madam, of another kind of "ideas," by your author named "bewitching or enchanting ideas," which are for the most part found in women, against which I cannot but take exception in the behalf of our sex; for, says he, "Women stamp ideas on themselves, whereby they, [244] no otherwise than witches driven about with a malignant spirit of despair, are oftentimes governed or snatched away unto those things, which otherwise they would not, and do bewail unto us their own and involuntary madness. These ideas are hurtful to themselves, and do, as it were, enchant, infatuate, and weaken themselves [. . .]."[19]

By this it appears that your author has never been in love, or else he would have found that men have as well bewitching ideas as women, and they are as hurtful to men as to women. Neither can I be persuaded to believe that men should not have as well mad ideas as women; for to mention no other example, some (I will not speak of your author), their writings and strange opinions in philosophy do sufficiently witness it; but whence those ideas do proceed, whether from the bridebed of the soul or the spleen, your author does not declare [. . .].

[. . .] To conclude, Madam, all these rehearsed opinions of your author concerning the grounds or principles of natural philosophy, if you desire my unfeigned judgment, I can say no more but that they show more fancy than reason and truth, and so do many others; and perhaps my opinions may be as far from truth as his, although their ground is sense and reason; for there is [245] no single creature in nature that is able to know the perfectest truth. But some opinions, to human sense and reason, may have more probability than others, and everyone thinks his to be most probable, according to his own fancy and imagination, and so I think of mine; nevertheless, I leave them to the censure of those that are endued with solid judgment and reason, and know how to discern between things of fancy and reason, and amongst the rest, I submit them to the

---

19. Van Helmont, *Oriatrike*, chapter 82, "Of Things Conceived, or Conceptions."

censure of your Ladyship. Whose solid and wise judgment is the rule of all the actions of, Madam,

Your faithful friend and servant.

3.

Madam,

Your author relating how he dissents from the "false doctrines," as he terms it, "of the schools," concerning the elements and their mixtures, qualities, temperaments, discords, etc., in order to[20] diseases, is pleased to say as follows: "I have sufficiently demonstrated that there are not four elements in nature, and by consequence, if there are only three, that four cannot go together or encounter; and that the fruits which antiquity has believed to be mixed bodies, and those composed from a concurrence of four elements, are [246] materially of only one element; also that those three elements are naturally cold [. . .]."[21]

To give you my opinion hereof, first, I think it too great a presumption in any man to feign himself so much above all the rest as to accuse all others of ignorance, and that none but he alone has the true knowledge of all things as infallible and undeniable, and that so many learned, wise, and ingenious men in so many ages have been blinded with errors; for certainly, no particular creature in nature can have any exact or perfect knowledge of natural things, and therefore opinions cannot be infallible truths, although they may seem probable; for how is it possible that a single finite creature should know the numberless varieties and hidden actions of nature?

Wherefore your author cannot say that he has demonstrated anything which could not be as much contradicted, and perhaps with more reason, than he has brought proofs and demonstrations. And thus when he speaks of elements, that there are not four in nature, and that they cannot go together or encounter, it may be his opinion; but others have brought as many reasons to the contrary, and I think with more probability; so as it is unnecessary to make a tedious

---

20. In order to: with respect to.

21. Van Helmont, *Oriatrike*, "A Passive Deceiving of the Schools of the Humorists," chapter 1, "That the Four Humors of the Galenists are Feigned."

discourse hereof, and therefore I'll refer you to those that have treated of it more learnedly and solidly than I can do.

But I perceive your author is [247] so much for art, and since he can make solid bodies liquid, and liquid bodies solid, he believes that all bodies are composed out of the element of water, and that water therefore is the first principle of all things; when as water, in my opinion, is but an effect, as all other natural creatures, and therefore cannot be a cause or principle of them. [. . .]

[248] [. . .] Lastly, as for his opinion that "there are no contraries in nature,"[22] I believe not in the essence or nature of matter; but sense and reason inform us that there are contraries in nature's actions, which are corporeal motions, which cause mixtures, qualities, degrees, discords, as also harmonious conjunctions and concords, compositions, divisions, and the like effects whatsoever.

But though your author seems to be an enemy to the mixtures of elements, yet he makes such a mixture of divinity and natural philosophy that all his philosophy is nothing but a mere hodge-podge, spoiling one with the other. And so I will leave it to those that delight in it, resting, Madam,

Your faithful friend and servant.

[275]   **11.**
_____

Madam,

You will cease to wonder that I am not altogether capable to understand your author's opinions in natural philosophy when you do but consider that his expressions are for the most part so obscure, mystical, and intricate as may puzzle any brain that has not the like genius or the same conceptions with your author [. . .].

[277] [. . .] However, Madam, I do not deny ideas, images, or conceptions of things, but I deny them only to be such powerful beings and principal efficient causes of natural effects; especially they being to your author neither bodies nor substances themselves. And as for the figure of a cherry, which your author

_____

22. Van Helmont, *Oriatrike*, "A Passive Deceiving of the Schools of the Humorists," chapter 1.

makes so frequent a repetition of, made by a longing woman on her child,[23] I daresay that there have been millions of women which have longed for some or other thing, and have not been satisfied with their desires, and yet their children have never had on their bodies the prints or marks of those things they longed for. But because some such figures are sometimes made by the irregular motions of animate matter, would this be a sufficient proof that all conceptions, ideas, and images have the like effects, after the same manner, by piercing or penetrating each other, and sealing or printing such or such a figure upon the body of the child? [. . .]

[278] [. . .] But, Madam, I have neither such an archeus, which can produce in my mind an idea of consent or approbation of these your author's opinions, nor such a light that is able to produce a beam of patience to tarry any longer upon the examination of them. Wherefore I beg your leave to cut off my discourse here, and only to subscribe myself, as I really am, Madam,

> Your humble and faithful servant.

12.

Madam,

[279] [. . .] And as for your author's opinion that "there are no contraries in nature," I am quite of a contrary mind, *viz.*, that there is a perpetual war and discord amongst the parts of nature, although not in the nature and substance of infinite matter, which is [280] of a simple kind and knows no contraries in itself, but lives in peace, when as the several actions are opposing and crossing each other; and truly, I do not believe that there is any part or creature of nature that has not met with opposers, let it be never so small or great. But as war is made by the division of nature's parts and variety of natural actions, so peace is caused by the unity and simplicity of the nature and essence of only matter, which nature is peaceable, being always one and the same, and having nothing in itself to be crossed or opposed by; when as the actions of nature or natural matter are continually striving against each other, as being various and different. [. . .]

---

23. Van Helmont suggests that a pregnant woman who craves cherries may give birth to a child with a cherry-shaped birthmark. See *Oriatrike*, chapter 83, "The Magnetic or Attractive Power or Faculty." Also see the Introduction (p. xli).

[281] [. . .] But your author, being a chemist, is much for the art of fire, although it is impossible for art to work as nature does; for art makes of natural creatures artificial monsters, and does oftener obscure and disturb nature's ordinary actions than prove any truth in nature. But nature, loving variety, does rather smile at art's follies than that she should be angry with her curiosity. Like as for example, a poet will smile in expressing the part or action of a fool.

Wherefore pure natural philosophers shall by natural sense and reason trace nature's ways and observe her actions more readily than chemists can do by fire and furnaces; for fire and furnaces do often delude the reason, blind the understanding, and make the judgment stagger. Nevertheless, your author is so taken with fire that from thence he imagines a formal light, which he believes to be the tip-top of life; but certainly he had, in my opinion, not so much light as to observe that all sorts of light are but creatures and not creators; for he judges several parts of matter as if they were several kinds of matter, which causes him often to err, although he conceits himself without any error. In which conceit I leave him, and rest, Madam,

Your faithful friend and humble servant.

[282]   **13.**

Madam,

The art of fire, I perceive, is in greater esteem and respect with your author than nature herself. For he says that "some things can be done by art, which nature cannot do";[24] nay, he calls art "the mistress of nature,"[25] and subjects whole nature unto chemical speculation; for "nothing," says he, "does more fully bring a man that is greedy of knowing to the knowledge of all things knowable than the fire"; "for the root or radical[26] knowledge of natural things consists in the fire."[27] "It pierces out the secrets of nature, and causes a further searching

---

24. Van Helmont, *Oriatrike*, chapter 11, "The Essay of the Meteor."

25. Van Helmont, *Oriatrike*, chapter 27, "Heat Doth Not Digest Efficiently, but Only Excitingly or By Way of Stirring Up."

26. Radical: fundamental.

27. Van Helmont, *Oriatrike*, chapter 7, "The Ignorant Natural Philosophy of Aristotle and Galen."

out in nature than all other sciences being put together; and pierces even into the utmost depths of real truth."[28] "It creates things which never were before."[29]

These and many more the like expressions he has in praise of chemistry. And truly, Madam, I cannot blame your author for commending this art, because it was his own profession, and no man will be so unwise as to dispraise his own art which he professes; but whether those praises and commendations do not exceed truth and express more than the art of fire can perform, I will let those judge that have more knowledge therein than I. But this I may say, that what art or science soever is in nature, let it be the chief of all, yet it can never be called the "mistress of nature," nor be said to perform more than nature does, except it be by a divine and supernatural [283] power; much less to create things which never were before, for this is an action which only belongs to God. The truth is, art is but a particular effect of nature, and as it were, nature's mimic or fool, in whose playing actions she sometimes takes delight. [. . .] I am sure that the art of fire cannot create and produce so as nature does, nor dissolve substances so as she does, nor transform and transchange as she does, nor do any effect like nature. And therefore I cannot so much admire this art as others do, for it appears to me to be rather a troubler than an assistant to nature, producing more monsters than perfect creatures; nay, it rather does shut the gates of truth than unlock the gates of nature. For how can art inform us of nature when as it is but an effect of nature?

You may say, the cause cannot be better known than by its effect; for the knowledge of the effect leads us to the knowledge of the cause. I answer, 'tis true. But you will consider that nature is an infinite cause and has infinite effects; and if you knew all the infinite effects in nature, then perhaps you might come to some knowledge of the cause; but to know nature by one single effect, as art is, is impossible; nay, no man knows this particular effect as yet perfectly. For who is he that [284] has studied the art of fire so as to produce all that this art may be able to afford? Witness the philosopher's stone.[30] Besides, how is it possible to find out the only cause by so numerous variations of the effects?

Wherefore it is more easy, in my opinion, to know the various effects in nature by studying the prime cause, than by the uncertain study of the inconstant effects to arrive to the true knowledge of the prime cause; truly it is much easier to walk in a labyrinth without a guide than to gain a certain knowledge

---

28. Van Helmont, *Oriatrike*, chapter 59, "A Modern Pharmacopolion and Dispensatory."

29. Van Helmont, *Oriatrike*, chapter 60, "The Power of Medicines."

30. On the philosopher's stone, see the Introduction (p. xxxix).

in any one art or natural effect, without nature herself be the guide, for nature is the only mistress and cause of all, which, as she has made all other effects, so she has also made arts for variety's sake; but most men study chemistry more for employment than for profit; not but that I believe there may be some excellent medicines found out and made by the art, but the expense and labor is more than the benefit; neither are all those medicines sure and certain, nor in all diseases safe; neither can this art produce so many medicines as there are several diseases in nature; and for the universal medicine and the philosopher's stone or elixir which chemists brag of so much, it consists rather in hope and expectation than in assurance; for could chemists find it out, they would not be so poor as most commonly they are, but richer than Solomon was, or any prince in the world, and might have done many famous acts with the supply of their vast golden treasures, to the eternal and immortal fame of their art; nay, gold being the idol of this world, they would be worshipped as well for the sake of gold as for their splendorous art; but how many endeavored and labored in vain and [285] without any effect?

"Gold is easier to be made than to be destroyed," says your author,[31] but I believe one is as difficult or impossible, nay more, than the other; for there is more probability of dissolving or destroying a natural effect by art than of generating or producing one; for art cannot go beyond her sphere of activity, she can but produce an artificial effect, and gold is a natural creature; neither were it justice that a particular creature of nature should have as much power to act or work as nature herself; but because neither reason nor art has found out as yet such a powerful opposite to gold as can alter its nature, men therefore conclude that it cannot be done. [. . .]

But leaving gold, since it is but a vain wish, I do verily believe that some of the chemical medicines do, in some desperate cases, many times produce more powerful and sudden effects than the [286] medicines of Galenists;[32] and therefore I do not absolutely condemn the art of fire, as if I were an enemy to it; but I am of an opinion that my opinions in philosophy, if well understood, will rather give a light to that art than obscure its worth; for if chemists did but study well the corporeal motions or actions of nature's substantial body, they would, by their observations, understand nature better than they do by the observation of the

31. Van Helmont, *Oriatrike*, chapter 55, "That the First Three Principles of the Chemists, nor the Essences of the Same, Are Not, or Do Not Belong unto the Army of Diseases."

32. Most trained physicians of mid-seventeenth-century England adhered to theories associated with the physician Galen (129–c. 216). On Galenism versus iatrochemical medical theories, see the Introduction (pp. xxxviii–xxxix).

actions of their art; and out of this consideration and respect, I should almost have an ambition to become an artist in chemistry, were I not too lazy and tender for that employment; but should I quit the one and venture the other, I am so vain as to persuade myself I might perform things worthy [of] my labor upon the ground of my own philosophy, which is substantial life, sense, and reason; for I would not study salt, sulfur, and mercury, but the natural motions of every creature, and observe the variety of nature's actions.

But, perchance, you will smile at my vain conceit,[33] and, it may be, I myself should repent of my pains unsuccessfully bestowed, my time vainly spent, my health rashly endangered, and my noble lord's estate unprofitably wasted in fruitless trials and experiments. Wherefore you may be sure that I will consider well before I act; for I would not lose health, wealth, and fame, and do no more than others have done, which truly is not much, their effects being of less weight than their words. But in the meantime, my study shall be bent to your service, and how to express myself worthily, Madam,

Your Ladyship's humble and faithful servant.

[289] **15.**

Madam,

Concerning sympathy and antipathy, and attractive or magnetic inclinations, which some do ascribe to the influence of the stars, others to an unknown spirit as the mover, others to the instinct of nature, hidden proprieties, and certain formal virtues [. . .]. If you please to have my opinion thereof, I think they are nothing else but plain passions and appetites. As for example, I take sympathy, as also magnetism or attractive power, to be such agreeable motions in one part or creature as to cause a fancy, love, and desire to some other part or creature; and antipathy, when these motions are disagreeable and produce contrary effects, as dislike, hate, and aversion to some part or creature. And as there are many sorts of such motions, so there are [290] many sorts of sympathies and antipathies, or attractions and aversions, made several manners or ways. For in some subjects, sympathy requires a certain distance; as for example in iron and

---

33. Conceit: notion.

the lodestone;[34] for if the iron be too far off, the lodestone cannot exercise its power, when as in other subjects, there is no need of any such certain distance, as between the needle and the north pole, as also the weapon-salve;[35] for the needle will turn itself towards the north whether it be near or far off from the north pole; and so, be the weapon which inflicted the wound never so far from the wounded person, as they say, yet it will nevertheless do its effect. But yet there must withal be some conjunction with the blood; for as your author mentions, the weapon shall be in vain anointed with the unguent unless it be made bloody, and the same blood be first dried on the same weapon.[36] [. . .]

[291] [. . .] And thus some sympathy and antipathy is made by a close conjunction or corporeal uniting of parts, but not all; neither is it required that all sympathy and antipathy must be mutual, or equally in both parties, so that that part or party which has a sympathetical affection or inclination to the other must needs receive the like sympathetical affection from that part again; for one man may have a sympathetical affection to another man, when as this man has an antipathetical aversion to him; and the same may be, for aught we know, between iron and the lodestone, as also between the needle and the north; for the needle may have a sympathy towards the north, but not again the north towards the needle; and so may the iron have towards the lodestone, but not again the lodestone towards the iron. Neither is sympathy or antipathy made by the issuing out of any invisible rays, for then the rays between the north and the needle would have a great way to reach. But a sympathetical inclination in a man towards another is made either by sight or hearing, either present or absent. The like of infectious diseases.

I grant that if both parties do mutually affect each other, and their motions be equally agreeable, then the sympathy is the stronger and will last the longer, and then there is a union, likeness, or [292] conformableness of their actions, appetites, and passions. For this kind of sympathy works no other effects but a conforming of the actions of one party to the actions of the other, as by way of imitation, proceeding from an internal sympathetical love and desire to please; for sympathy does not produce an effect really different from itself, or else the sympathy between iron and the lodestone would produce a third creature different from themselves, and so it would do all other creatures.

---

34. Lodestone: a naturally magnetized iron oxide.

35. On the weapon-salve and the dispute over its efficacy, see the Introduction (p. xxxvii).

36. Van Helmont, *Oriatrike*, chapter 112, "The Magnetic or Attracting Cure of Wounds."

But as I mentioned above, there are many sorts of attractions in nature, and many several and various attractions in one sort of creatures, nay, so many in one particular as not to be numbered; for there are many desires, passions, and appetites which draw or entice a man to something or other, as for example, to beauty, novelty, luxury, covetousness, and all kinds of virtues and vices; and there are many particular objects not in every one of these, as for example, in novelty. For there are so many desires to novelty as there are senses, and so many novelties that satisfy those desires, as a novelty to the ear, a novelty to the sight, to touch, taste, and smell; besides, in every one of these, there are many several objects. To mention only one example, for the novelty of sight, I have seen an ape, dressed like a cavalier and riding on horseback with his sword by his side, draw a far greater multitude of people after him than a lodestone of the same bigness of the ape would have drawn iron; and as the ape turned, so did the people, just like as the needle turns to the north; and this is but one object in one kind of attraction, *viz.*, novelty. But there be [293] millions of objects besides. In like manner good cheer draws abundance of people, as is evident, and needs no demonstration.

Wherefore, as I said in the beginning, sympathy is nothing else but natural passions and appetites, as love, desire, fancy, hunger, thirst, etc., and its effects are concord, unity, nourishment, and the like. But antipathy is dislike, hate, fear, anger, revenge, aversion, jealousy, etc., and its effects are discord, division, and the like. And such an antipathy is between a wolf and a sheep, a hound and a hare, a hawk and a partridge, etc., For this antipathy is nothing else but fear in the sheep to run away from the wolf, in the hare to run from the hound, in the partridge to fly from the hawk; for life has an antipathy to that which is named death; and the wolf's stomach has a sympathy to food, which causes him to draw near or run after those creatures he has a mind to feed on.

But you will say, some creatures will fight and kill each other not for food, but only out of an antipathetical nature. I answer, when as creatures fight and endeavor to destroy each other, if it be not out of necessity, as to preserve and defend themselves from hurt or danger, then it is out of revenge, or anger, or ambition, or jealousy, or custom of quarrelling, or breeding. As for example cocks of the game that are bred to fight with each other, and many other creatures as bucks, stags, and the like, as also birds, will fight as well as men, and seek to destroy each other through jealousy; when as, had they no females amongst them, they would perhaps live quiet enough, rather as sympathetical friends than [294] antipathetical foes; and all such quarrels proceed from a sympathy to their own interest.

But you may ask me what the reason is that some creatures, as for example, mankind, some of them, will not only like one sort of meat better than another of equal goodness and nourishment, but will like and prefer sometimes a worse sort of meat before the best, to wit, such as has neither a good taste nor nourishment? I answer, this is nothing else but a particular, and most commonly an inconstant appetite; for after much eating of that they like best, especially if they get a surfeit,[37] their appetite is changed to aversion; for then all the feeding motions and parts have as much, if not more antipathy to those meats, as before they had a sympathy to them.

Again, you may ask me the reason why a man, seeing two persons together which are strangers to him, does affect[38] one better than the other; nay if one of these persons be deformed or ill-favored, and the other well-shaped and handsome, yet it may chance that the deformed person shall be more acceptable in the affections and eyes of the beholder than he that is handsome. I answer, there is no creature so deformed but has some agreeable and attractive parts, unless it be a monster, which is never loved but for its rarity and novelty, and nature is many times pleased with changes, taking delight in variety. And the proof that such a sympathetical affection proceeds from some agreeableness of parts is that if those persons were veiled, there would not proceed such a partial choice or judgment from any to them.

You may ask me further whether passion and appetite are also the cause of the sympathy which is in the lodestone towards iron, and in the [295] needle towards the north? I answer, yes. For it is either for nourishment, or refreshment, or love and desire of association, or the like, that the lodestone draws iron, and the needle turns towards the north. [. . .]

[297] [. . .] And thus, to shut up my discourse, I repeat again that sympathy and antipathy are nothing else but ordinary passions and appetites amongst several creatures, which passions are made by the rational animate matter, and the appetites by the sensitive, both giving such or such motions to such or such creatures; for cross motions in appetites and passions make antipathy, and agreeable motions in appetites and passions make sympathy, although the creatures be different wherein these motions, passions, and appetites are made; and as without an object a pattern cannot be, so without inherent or natural passions and appetites there can be no sympathy or antipathy. And there being also such sympathy between your Ladyship and me, I think myself the happiest creature

---

37. Surfeit: overindulgence, excess.
38. Affect: prefer.

for it, and shall make it my whole study to imitate your Ladyship and conform all my actions to the rule and pattern of yours, as becomes, Madam,

Your Ladyship's faithful friend and humble servant.

[298]   **16.**

Madam,

My opinion of witches and witchcraft (of whose power and strange effects your author is pleased to relate many stories) in brief, is this: my sense and reason do inform me that there is natural witchcraft, as I may call it, which is sympathy, antipathy, magnetism, and the like, which are made by the sensitive and rational motions between several creatures, as by imagination, fancy, love, aversion, and many the like; but these motions, being sometimes unusual and strange to us, we not knowing their causes (for what creature knows all motions in nature, and their ways?) do stand amazed at their working power; and by reason we cannot assign any natural cause for them, are apt to ascribe their effects to the devil; but that there should be any such devilish witchcraft, which is made by a covenant and agreement with the devil, by whose power men do enchant or bewitch other creatures, I cannot readily believe.

Certainly, I dare say that many a good, old honest woman has been condemned innocently, and suffered death wrongfully, by the sentence of some foolish and cruel judges, merely upon this suspicion of witchcraft, when as really there has been no such thing, for many things are done by sleights or juggling arts,[39] wherein neither the devil nor witches are actors. [. . .]

[299] [. . .] But, as I said before, I believe there is natural magic; which is, that the sensitive and rational matter oft moves such a way as is unknown to us; and in the number of these is also the bleeding of a murdered body at the presence of the murderer, which your author mentions;[40] for the corporeal motions in the murdered body may move so as to work such effects which are more than ordinary; for the animal figure being not so quickly dissolved, the animal motions are not so soon altered (for the dissolving of the figure is nothing else but an alteration of its motions); and this dissolution is not done in an instant of time,

---

39. Juggling arts: conjuring, playing tricks.
40. Van Helmont, *Oriatrike*, chapter 112, "Of the Magnetic or Attracting Cure of Wounds."

but by degrees. But yet I must confess, it is not a common action in nature, for nature has both common, and singular or particular actions. As for example, madness, natural folly, and many the like are but in some particular persons; for if those actions were general and common, then all [300] or most men would be either mad or fools, but though there are too many already, yet all men are not so; and so some murdered bodies may bleed or express some alterations at the presence of the murderer, but I do not believe that all do so; for surely in many, not any alteration will be perceived, and others will have the same alterations without the presence of the murderer. [. . .]

[301] [. . .] But to return to magnetism; I am absolutely of opinion that it is naturally effected by natural means, without the concurrence of immaterial spirits either good or bad, merely by natural corporeal sensitive and rational motions; and, for the most part, there must be a due approach between the agent and the patient, or otherwise the effect will hardly follow, as you may see by the lodestone and iron. Neither is the influence of the stars performed beyond a certain distance, that is, such a distance as is beyond sight or their natural power to [302] work; for if their light comes to our eyes, I know no reason against it, but their effects may come to our bodies.

And as for infectious diseases, they come by a corporeal imitation, as by touch, either of the infected air, drawn in by breath, or entering through the pores of the body, or of some things brought from infected places, or else by hearing, but diseases caused by conceit have their beginning, as all alterations have, from the sensitive and rational motions, which do not only make the fear and conceit, but also the disease; for as a fright will sometimes cure diseases, so it will sometimes cause diseases; but as I said, both fright, cure, and the disease are made by the rational and sensitive corporeal motions within the body, and not by supernatural magic, as satanical witchcraft, entering from without into the body by spiritual rays.

But having discoursed hereof in my former letter, I will not trouble you with an unnecessary repetition thereof; I conclude therefore with what I began, *viz.*, that I believe natural magic to be natural corporeal motions in natural bodies. Not that I say nature in herself is a magicianess, but it may be called natural magic or witchcraft merely in respect to our ignorance; for though nature is old, yet she is not a witch, but a grave, wise, methodical matron, ordering her infinite family, which are her several parts, with ease and facility, without needless troubles and difficulties; for these are only made through the ignorance of her several parts or particular creatures, not understanding their mistress, nature, and her actions and government, for which they cannot be blamed; for how should a part [303] understand the infinite body when it does not understand itself; but nature understands her parts better than they do her. And so leaving wise nature and the ignorance of her particulars, I understand myself so far, that I am, Madam,

Your humble and faithful servant.

17.

Madam,

I am not of your author's opinion that "time has no relation to motion, but that time and motion are as unlike and different from each other as finite from infinite, and that it has its own essence or being, immoveable, unchangeable, individable, and unmixed with things, nay, that time is plainly the same with eternity."[41]

For, in my opinion, there can be no such thing as time in nature, but what man calls "time" is only the variation of natural motions; wherefore time and the alteration of motion is one and the same thing under two different names; and as matter, figure, and motion are inseparable, so is time inseparably united, or rather the same thing with them, and not a thing subsisting by itself; and as long as matter, motion, and figure have been existent, so long has time; and as long as they last, so long does time.

[304] But when I say time is the variation of motion, I do not mean the motion of the sun or moon, which makes days, months, years, but the general motions or actions of nature, which are the ground of time; for were there no motion, there would be no time; and since matter is dividable and in parts, time is so too; neither has time any other relation to duration than what nature herself has. Wherefore your author is mistaken when he says motion is made in time, for motion makes time, or rather is one and the same with time; and succession is one and the same with time; and succession is no more a stranger to motion than motion is to nature, as being the action of nature, which is the eternal servant of God. [. . .] But time is not eternity, for eternity has no change, although your author makes time and eternity all one, and a being or substance itself. [. . .]

[305] [. . .] Your Ladyship's faithful friend and humble servant.

---

41. Van Helmont, *Oriatrike*, chapter 89, "Of Time."

18.

Madam,

[306] [...] But the mistake is that most men do not or will not conceive that there is a difference and variety of the corporeal sensitive and rational motions in every creature; but they imagine that if all creatures should have life, sense, and reason, they must of necessity have all alike the same motions, without any difference; and because they do not perceive the animal motions in a stone or tree, they are apt to deny to them all life, sense, and motion.

Truly, Madam, I think no man will be so mad or irrational as to say a stone is an animal, or an animal is a tree, because a stone and tree have sense, life, and motion; everybody knows that their natural figures are different, and if their natures be different, then they cannot have the same motions, for the corporeal motions [307] do make the nature of every particular creature, and their differences; and as the corporeal motions act, work, or move, so is the nature of every figure. Wherefore, nobody, I hope, will count me so senseless that I believe sense and life to be after the like manner in every particular creature or part of nature; as for example, that a stone or tree has animal motions and does see, touch, taste, smell, and hear by such sensitive organs as an animal does; but my opinion is that all sense is not bound up to the sensitive organs of an animal, nor reason to the kernel of a man's brain, or the orifice of the stomach, or the fourth ventricle of the brain, or only to a man's body; for though we do not see all creatures move in that manner as man or animals do, as to walk, run, leap, ride, etc., and perform exterior acts by various local motions; nevertheless, we cannot in reason say they are void and destitute of motions in nature. For what man knows the variety of motions in nature.

Do we not see that nature is active in everything, yea, the least of her creatures. For example, how some things do unanimously conspire and agree, others antipathetically flee from each other; and how some do increase, others decrease; some dissolve, some consist, and how all things are subject to perpetual changes and alterations; and do you think all this is done without motion, life, sense, and reason? I pray you consider, Madam, that there are internal motions as well as external, alterative as well as constitutive; and several other sorts of motions not perceptible by our senses, and therefore it is impossible that all creatures should move after one sort of motions.

But you will say, motion may be granted, but not life, sense, and reason. I [308] answer, I would fain know the reason why not; for I am confident that no man can in truth affirm the contrary. What is life, but sensitive motion? What is reason, but rational motion? And do you think, Madam, that anything can move itself without life, sense, and reason? I, for my part, cannot imagine it should; for it would neither know why, whither, nor what way or how to move.

But you may reply, motion may be granted, but not self-motion, and life, sense, and reason, do consist in self-motion. I answer, this is impossible; for not anything in nature can move naturally without natural motion, and all natural motion is self-motion. If you say it may be moved by another, my answer is, first, that if a thing has no motion in itself, but is moved by another which has self-motion, then it must give that immovable body motion of its own, or else it could not move, having no motion at all; for it must move by the power of motion, which is certain; and then it must move either by its own motion or by a communicated or imparted motion; if by a communicated motion, then the self-moveable thing or body must transfer its own motion into the immoveable, and lose so much of its own motion as it gives away, which is impossible, as I have declared heretofore at large, unless it do also transfer its moving parts together with it, for motion cannot be transferred without substance. But experience and observation witness the contrary.

Next, I say, if it were possible that one body did move another, then most part of natural creatures, which are counted immoveable of themselves, or inanimate and destitute of self-motion, must be moved by a forced or violent, and not by a natural motion; for all motion that proceeds [309] from an external agent or moving power is not natural, but forced, only self-motion is natural; and then one thing moving another in this manner, we must at last proceed to such a thing which is not moved by another, but has motion in itself, and moves all others; and perhaps, since man and the rest of animals have self-motion, it might be said that the motions of all other inanimate creatures, as they call them, do proceed from them; but man, being so proud, ambitious, and self-conceited, would soon exclude all other animals and ascribe this power only to himself, especially since he thinks himself only endued with reason, and to have this prerogative above all the rest, as to be the sole rational creature in the world. Thus you see, Madam, what confusion, absurdity, and constrained work will follow from the opinion of denying self-motion, and so consequently, life and sense to natural creatures. [. . .]

[310] [. . .] But to conclude, the motions of the exterior agent and the motions of the patient[42] do sometimes join and unite, as in one action to one effect,

---

42. By "patient," Cavendish means anything that is acted upon by something else (the "agent").

and sometimes the motions of the agent are only an occasion, but not a co-workman in the production of [311] such an effect, as the motions of the patient do work; neither can the motions of the agent work totally and merely of themselves such or such effects, without the assistance or concurrence of the motions of the patient, but the motions of the patient can. And there is nothing that can prove more evidently that matter moves itself, and that exterior agents or bodies are only an occasion to such or such a motion in another body, than to see how several things put into one and the same fire do alter after several modes; which shows it is not the only action of fire, but the interior motions of the body thrown into the fire, which do alter its exterior form or figure. And thus, I think I have said enough to make my opinions clear, that they may be the better understood; which is the only aim and desire of, Madam,

Your humble and faithful servant.

19.

Madam,

Your author is not a natural, but a divine philosopher, for in many places he undertakes to interpret the Scripture; wherein, to my judgment, he expresses very strange opinions [. . .].

[313] [. . .] But your author shows a great affection to the female sex when he says that God does love women before men, and that he has given them a free gift of devotion before men;[43] when as others do lay all the fault upon the woman, that she did seduce the man;[44] however, in expressing his affection for women, your author expresses a partiality in God. And as for his opinion that God creates more daughters than males, and that more males are extinguished by diseases, travels, wars, duels, shipwrecks, and the like; [314] truly, I am of the same mind that more men are killed by travels, war, duels, shipwrecks, etc. than women; for women never undergo these dangers, neither do so many kill themselves with intemperate drinking as men do; but yet I believe that death is as general, and not more favorable to women than he is to men; for though women be not slain in wars like men (although many are, by the cruelty of men, who not

---

43. Van Helmont, *Oriatrike*, chapter 93, "The Position is Demonstrated."
44. I.e., Eve and Adam.

regarding the weakness of their sex, do inhumanely kill them), yet many do die in child-bed, which is a punishment only concerning the female sex. [. . .]

[. . .] There are many more such expressions in your author's works, which, in my opinion, do rather detract from the greatness of the omnipotent God than manifest his glory. As for example that man is the clothing of the Deity and the sheath of the kingdom of God, and many the like. Which do not belong to God; for God is beyond all [315] expression, because he is infinite; and when we name God, we name an inexpressible and incomprehensible being; and yet we think we honor God when we express him after the manner of corporeal creatures.

Surely, the noblest creature that ever is in the world is not able to be compared to the most glorious God, but whatsoever comparison is made detracts from his glory. And this, in my opinion, is the reason that God forbade any likeness to be made of him, either in heaven or upon earth, because he exceeds all that we might compare or liken to him. And as men ought to have a care of such similizing expressions, so they ought to be careful in making interpretations of the Scripture and expressing more than the Scripture informs; for what is beyond the Scripture is man's own fancy, and to regulate the word of God after man's fancy, at least to make his fancy equal with the word of God, is irreligious.

Wherefore, men ought to submit, and not to pretend to the knowledge of God's counsels and designs above what he himself has been pleased to reveal. As for example, to describe of what figure God is, and to comment and descant upon the articles of faith; as how man was created; and what he did in the state of innocence; how he did fall; and what he did after his fall; and so upon the rest of the articles of our creed, more than the Scripture expresses, or is conformable to it. For if we do this, we shall make a romance of the holy Scripture with our paraphrastical[45] descriptions. Which alas! is too common already.

The truth is, natural philosophers should only contain themselves within the sphere of nature, and not trespass upon the revelation of the Scripture, [316] but leave this profession to those to whom it properly belongs. I am confident, a physician or any other man of a certain profession would not take it well if others who are not professed in that art should take upon them to practice the same. And I do wonder why everybody is so forward to encroach upon the holy profession of divines, which yet is a greater presumption than if they did it upon any other; for it contains not only a most hidden and mystical knowledge, as treating of the highest subject, which is the most glorious and incomprehensible

---

45. Paraphrastical: paraphrased.

God and the salvation of our souls; but it is also most dangerous, if not interpreted according to the Holy Spirit, but to the bias of men's fancy.

Wherefore, Madam, I am afraid to meddle with divinity in the least thing,
lest I incur the hazard of offending the divine truth, and spoil the excellent art
of philosophizing; for a philosophical liberty and a supernatural faith are two
different things, and suffer no commixture; as I have declared sufficiently heretofore.[46] And this you will find as much truth, as that I am, Madam,

Your constant friend and faithful servant.

[317]   **20.**

Madam,

Although your author is of the opinion of Plato in making "three sorts of
atheists: one that believes no gods; another, which indeed admits of gods, yet
such as are uncareful of us and despisers of small matters, and therefore also
ignorant of us; and lastly, a third sort, which although they believe the gods to
be expert in the least matters, yet do suppose that they are flexible and indulgent
toward the smallest cold prayers or petitions,"[47] yet I cannot approve of this
distinction, for I do understand but one sort of atheists; that is, those which
believe no God at all; but those that believe there is a God, although they do not
worship him truly, nor live piously and religiously as they ought, cannot in truth
be called atheists, or else there would be innumerable sorts of atheists; to wit,
all those that are either no Christians or not of this or that opinion in Christian
religion, besides all them that live wickedly, impiously, and irreligiously; for to
know and be convinced in his reason that there is a God, and to worship him
truly, according to his holy precepts and commands, are two several things.

And as for the first, that is, for the rational knowledge of the existence of
God, I cannot be persuaded to believe there is any man which has sense and
reason that does not acknowledge a God; nay, I am sure there is no part of

---

46. In this volume, see, for example, section 1, letter 1, and section 2, letter 29.
47. Van Helmont, *Oriatrike*, chapter 35, "Of the Image of the Mind." Book 10 of Plato's *Laws*
distinguishes between those who believe (1) in no gods at all; (2) in gods that take no interest in
human affairs; and (3) in gods that do take an interest in human affairs, but do not dispense justice
because they accept bribes from the wicked.

nature which is void and destitute of this knowledge of [318] the existence of an infinite, eternal, immortal, and incomprehensible Deity; for every creature, being endued with sense and reason and with sensitive and rational knowledge, there can no knowledge be more universal than the knowledge of a God, as being the root of all knowledge. And as all creatures have a natural knowledge of the infinite God, so it is probable they worship, adore, and praise his infinite power and bounty, each after its own manner and according to its nature; for I cannot believe God should make so many kinds of creatures, and not be worshipped and adored but only by man.

Nature is God's servant, and she knows God better than any particular creature; but nature is an infinite body, consisting of infinite parts, and if she adores and worships God, her infinite parts, which are natural creatures, must of necessity do the like, each according to the knowledge it has; but man in this particular goes beyond others, as having not only a natural but also a revealed knowledge of the most holy God; for he knows God's will not only by the light of nature but also by revelation, and so more than other creatures do, whose knowledge of God is merely natural. But this revealed knowledge makes most men so presumptuous that they will not be content with it, but search more and more into the hidden mysteries of the incomprehensible Deity, and pretend to know God as perfectly, almost, as themselves; describing his nature and essence, his attributes, his counsels, his actions, according to the revelation of God (as they pretend), when as it is according to their own fancies. So proud and [319] presumptuous are many. But they show thereby rather their weaknesses and follies than any truth; and all their strict and narrow pryings into the secrets of God are rather unprofitable, vain, and impious, than that they should benefit either themselves or their neighbor; for do all we can, God will not be perfectly known by any creature. The truth is, it is a mere impossibility for a finite creature to have a perfect idea of an infinite being, as God is; be his reason never so acute or sharp, yet he cannot penetrate what is impenetrable, nor comprehend what is incomprehensible.

Wherefore, in my opinion, the best way is humbly to adore what we cannot conceive, and believe as much as God has been pleased to reveal without any further search; lest we, diving too deep, be swallowed up in the bottomless depth of his infiniteness. Which I wish everyone may observe, for the benefit of his own self and of others, to spend his time in more profitable studies than vainly to seek for what cannot be found. And with this hearty wish I conclude, resting, Madam,

Your faithful friend and servant.

Madam,

Your author is so much for spirits that he does not stick[48] to affirm "that bodies scarce make up a moiety or half part of the world; but spirits, even by themselves, have or possess their moiety, and indeed the whole world."[49] If he mean bodiless and incorporeal spirits, I cannot conceive how spirits can take up any place, for place belongs only to body, or a corporeal substance, and millions of immaterial spirits, nay, were their number infinite, cannot possess so much place as a small pinpoint, for incorporeal spirits possess no place at all. Which is the reason that immaterial and a material infinite cannot hinder, oppose, or obstruct each other; and such an infinite, immaterial spirit is God alone.

But as for created immaterial spirits, as they call them, it may be questioned whether they be immaterial or not; for there may be material spirits as well as immaterial, that is, such pure, subtle, and agile substances as cannot be subject to any human sense, which may be purer and subtler than the most refined air or purest light; I call them material spirits, only for distinction's sake, although it is more proper to call them material substances. But be it that there are immaterial spirits, yet they are not natural, but supernatural; that is, not substantial parts of nature; for nature is material, or corporeal, and so are all her creatures, and whatsoever is not material is no part of [321] nature, neither does it belong any ways to nature. Wherefore, all that is called immaterial is a natural nothing, and an immaterial natural substance, in my opinion, is nonsense.

And if you contend with me that created spirits, as good and bad angels, as also the immortal mind of man, are immaterial, then I say they are supernatural; but if you say they are natural, then I answer they are material. And thus I do not deny the existence of immaterial spirits, but only that they are not parts of nature, but supernatural; for there may be many things above nature, and so above a natural understanding and knowledge, which may nevertheless have their being and existence, although they be not natural, that is, parts of nature. Neither do I deny that those supernatural creatures may be amongst natural creatures, that is, have their subsistence amongst them and in nature; but they are not so commixed with them as the several parts of

---

48. Stick: hesitate.
49. Van Helmont, *Oriatrike*, chapter 112, "The Magnetic or Attracting Cure of Wounds."

matter are, that is, they do not join to the constitution of a material creature; for no immaterial can make a material or contribute anything to the making or production of it; but such a commixture would breed a mere confusion in nature. Wherefore, it is quite another thing to be in nature, or to have its subsistence amongst natural creatures in a supernatural manner or way, and to be a part of nature. I allow the first to immaterial spirits, but not the second, *viz.*, to be parts of nature.

But what immaterial spirits are, both in their essence or nature and their essential properties, it being supernatural and above natural reason, I cannot determine anything thereof. Neither dare I say they are spirits like [322] God is, that is, of the same essence or nature, no more than I dare say or think that God is of a human shape or figure, or that the nature of God is as easy to be known as any notion else whatsoever, and that we may know as much of him as of anything else in the world. For if this were so, man would know God as well as he knows himself; but God and his attributes are not so easily known as man may know himself and his own natural proprieties; for God and his attributes are not conceivable or comprehensible by any human understanding, which is not only material but also finite; for though the parts of nature be infinite in number, yet each is finite in itself, that is, in its figure, and therefore no natural creature is capable to conceive what God is; for he being infinite, there is also required an infinite capacity so to conceive him; nay, nature herself, although she is infinite, yet cannot possibly have an exact notion of God, by reason she is material and God is immaterial; and if the infinite servant of God is not able to conceive God, much less will a finite part of nature do it. [. . .]

[323] [. . .] And as I do not meddle with any divine mysteries, but subject myself, concerning my faith or belief and the regulating of my actions for the obtaining of eternal life, wholly under the government and doctrine of the church, so I hope they will also grant me leave to have my liberty concerning the contemplation of nature and natural things, that I may discourse of them with such freedom as mere natural philosophers use, or at least ought, to do; and thus I shall be both a good Christian and a good natural philosopher. Unto which, to make the number perfect, I will add a third, which is, I shall be, Madam,

Your real and faithful friend and servant.

Madam,

Your author's comparison of the sun with the immaterial or divine soul in man[50] makes me almost of opinion that the sun is the soul of this world we inhabit, and that the fixed stars, which are counted suns by some, may be souls to some other worlds; for every one man has but one immaterial or divine soul, which is said to be individable and simple in its essence, and therefore unchangeable; and if the sun be like this immaterial soul, then the moon may be like the material soul.

But as for the production of this immaterial and divine soul in man, whether it come by an immediate creation from God or be derived by a successive propagation from parents upon their children, I cannot determine anything, being supernatural and not belonging to my study; nevertheless, the propagation from parents seems improbable to my reason; for I am not capable to imagine how an immaterial soul, being individable, should beget [330] another. Some may say, by imprinting or sealing, *viz.*, that the soul does print the image of its own figure upon the spirit of the seed; which if so, then first there will only be a production of the figure of the soul, but not of the substance, and so the child will have but the image of the soul, and not a real and substantial soul. Secondly, every child of the same parents would be just alike, without any distinguishment; if not in body, yet in the faculties and proprieties of their mind or souls. Thirdly, there must be two prints of the two souls of both parents upon one creature, to wit, the child; for both parents do contribute alike to the production of the child, and then the child would either have two souls, or both must be joined as into one; which how it can be, I am not able to conceive. Fourthly, if the parents print the image of their souls upon the child, then the child's soul bears not the image of God, but the image of man, to wit, his parents. Lastly, I cannot understand how an immaterial substance should make a print upon a corporeal substance, for printing is a corporeal action and belongs only to bodies.

Others may say that the soul is from the parents transmitted into the child, like as a beam of light; but then the souls of the parents must part with some of their own substance; for light is a substance dividable, in my opinion; and if it were not, yet the soul is a substance, and cannot be communicated without

---

50. Van Helmont, *Oriatrike*, chapter 37, "Of the Seat of the Soul," and chapter 35, "Of the Image of the Mind."

losing some of his own substance, but that is impossible; for the immaterial soul, being individable, cannot be diminished nor increased in its substance or nature.

Others again will have the soul produced by certain ideas; but ideas, being corporeal, cannot produce a substance incorporeal [331] or spiritual. Wherefore I cannot conceive how the souls of the parents, being individable in themselves, and not removable out of their bodies until the time of death, should commix so as to produce a third immaterial soul like to their own. You will say, as the sun, which is the fountain of heat and light, heats and enlightens, and produces other creatures. But I answer, the sun does not produce other suns, at least not to our knowledge.

'Tis true, there are various and several manners and ways of productions, but they are all natural, that is, material or corporeal; to wit, productions of some material beings or corporeal substances; but the immaterial soul not being in the number of these, it is not probable that she is produced by the way of corporeal productions, but created and infused from God, according to her nature, which is supernatural and divine. But being the image of God, how she can be defiled with the impurity of sin, and suffer eternal damnation for her wickedness, without any prejudice to her creator, I leave to the church to inform us thereof.

Only one question I will add, whether the soul be subject to sickness and pain? To which I answer, as for the supernatural and divine soul, although she be a substance, yet being not corporeal but spiritual, she can never suffer pain, sickness, nor death; but as for the natural soul, to speak properly, there is no such thing in nature as pain, sickness, or death; unless in respect to some particular creatures composed of natural matter; for what man calls "sickness," "pain," and "death" are nothing else but the motions of nature; for though there is but one only matter, that is, nothing but mere matter in [332] nature, without any commixture of either a spiritual substance, or anything else that is not matter, yet this mere matter is of several degrees and parts and is the body of nature. Besides, as there is but one only matter, so there is also but one only motion in nature, as I may call it, that is, mere corporeal motion, without any rest or cessation, which is the soul of that natural body, both being infinite; but yet this only corporeal motion is infinitely various in its degrees or manners, and ways of moving; for it is nothing else but the action of natural matter, which action must needs be infinite, being the action of an infinite body, making infinite figures and parts.

These motions and actions of nature, since they are so infinitely various, when men chance to observe some of their variety, they call them by some proper name to make a distinguishment, especially those motions which belong to the figure of their own kind; and therefore when they will express the motions of dissolution of their own figure, they call them "death"; when they will express the motions of production of their figure, they call them "conception" and "generation"; when they will express the motions proper for the consistence, continuance, and perfection of their figure, they call them "health"; but when they will express the motions contrary to these, they call them "sickness," "pain," "death," and the like. And hence comes also the difference between regular and irregular motions; for all those motions that belong to the particular nature and consistence of any figure, they call "regular," and those which are contrary to them, they call "irregular."

And thus you see, Madam, that there is no such thing in nature as death, sickness, pain, [333] health, etc., but only a variety and change of the corporeal motions, and that those words express nothing else but the variety of motions in nature; for men are apt to make more distinctions than nature does. Nature knows of nothing else but of corporeal figurative motions, when as men make a thousand distinctions of one thing, and confound and entangle themselves so, with beings, non-beings, and neutral-beings, corporeals and incorporeals, substances and accidents, or manners and modes of substances, new creations and annihilations, and the like, as neither they themselves nor anybody else is able to make any sense thereof; for they are like the tricks and sleights of jugglers, 'tis here, 'tis gone; and amongst those authors which I have read as yet, the most difficult to be understood is this author which I am now perusing, who runs such divisions and cuts nature into so small parts as the sight of my reason is not sharp enough to discern them. Wherefore I will leave them to those that are more quick-sighted than I, and rest, Madam,

Your constant friend and faithful servant.

[338]   **25.**

Madam,

[. . .] [340] [. . .] But your author, by his leave, confounds reason and reasoning, which are two several and distinct things; for reasoning and arguing differs as much from reason, as doubtfulness from certainty of knowledge, or a wavering mind from a constant mind; for reasoning is the discursive, and reason the understanding part in man, and therefore I can find no great difference between understanding and reason. Neither can I be persuaded that reason should not remain with man after this life and enter with him into heaven, although your author speaks much against it;[51] for if man shall be the same then which he is now in body, why not in soul also? 'Tis true, the Scripture says, he shall have a more glorious body, but it does not say that some parts of the body shall be cast away or remain behind; and if not of the body, why of the soul? Why shall reason, which is the chief part of the natural soul, be wanting? [. . .]

[341] [. . .] I grant that perfect truth requires not reasoning or arguing, as whether it be so or not; but yet it requires reason, as to confirm it to be so or not so. For reason is the confirmation of truth, and reasoning is but inquisition into truth. Wherefore, when our souls shall be in the fullness of blessedness, certainly they shall not be so dull and stupid, but observe distinctions between God, angels, and sanctified souls; as also, that our glory is above our merit, and that there is great difference between the damned and the blessed, and that God is an eternal and infinite being, and only to be adored, admired, and loved. All which the soul cannot know without the distinguishment of reason; otherwise we might say the souls in heaven love, joy, admire, and adore, but know not what, why, or wherefore. For shall the blessed souls present continual praises without reason? Have they not reason to remember the mercies of God and the merits of his son? For without remembrance of them, they cannot give a true acknowledgment, although your author says there is no use of memory or remembrance in heaven. But surely, I believe there is; for if there were not memory in heaven, the penitent thief upon the cross his prayers had been in vain; for he desired our savior to remember him when he did come into his kingdom. Wherefore if there be understanding in heaven, there is also reason; and if there be reason, there is memory also. For all souls in heaven, as [342] well as on earth, have reason to adore, love, and praise God.

But, Madam, my study is in natural philosophy, not in theology; and therefore I'll refer you to divines, and leave your author to his own fancy, who by his singular vision tells us more news of our souls than our savior did after his death and resurrection. Resting in the meantime, Madam,

Your faithful friend and servant.

---

51. Van Helmont, *Oriatrike*, chapter 3, "The Hunting and Searching Out of Sciences," and chapter 35, "Of the Image of the Mind."

26.

Madam,

Concerning those parts and chapters of your author's works which treat of physic,[52] before I begin to examine them, I beg leave of you in this present to make some reflections first upon his opinions concerning the nature of health and diseases. As for health, he is pleased to say that "it consists not in a just temperature of the body, but in a sound and entire life; for otherwise, a temperature of body is as yet in a dead carcass newly killed, where notwithstanding there is now death, but not life, not health."[53] Also he says "that no disease is in a dead carcass."[54] To which I answer that, in my opinion, life is in a dead carcass as well as in a living animal, although not such a life as that creature had before it became a carcass, and the [343] temperature of that creature is altered with the alteration of its particular life; for the temperature of that particular life which was before in the animal does not remain in the carcass in such a manner as it was when it had the life of such or such an animal; nevertheless, a dead carcass has life, and such a temperature of life, as is proper and belonging to its own figure; for there are as many different lives as there be different creatures, and each creature has its particular life and soul, as partaking of sensitive and rational matter. And if a dead carcass has life and such a temperature of motions as belong to its own life, then there is no question but these motions may move sometimes irregularly in a dead carcass as well as in any other creature; and since health and diseases are nothing else but the regularity or irregularity of sensitive corporeal motions, a dead carcass having irregular motions may be said as well to have diseases as a living body, as they name it, although it is no proper or usual term for others, but only for animals. [. . .]

[344] [. . .] But concerning diseases, your author's opinion is that a disease is as natural as health.[55] I answer, 'tis true, diseases are natural; but if we could find out the art of healing as well as the art of killing and destroying, and the art of uniting and composing, as well as the art of separating and dividing, it would be very beneficial to man; but this may easier be wished for than obtained; for

---

52. Physic: medicine.

53. Van Helmont, *Oriatrike*, chapter 55, "The Author Answers." Van Helmont's text says "sound or entire life" rather than "sound and entire life."

54. Van Helmont, *Oriatrike*, chapter 67, "The Subject of Inhering, of Diseases, is in the Point of Life."

55. Van Helmont, *Oriatrike*, chapter 68, "I Proceed Unto the Knowledge of Diseases." Van Helmont's claim is that diseases are "active beings, admitted into nature by natural principles."

nature, being a corporeal substance, has infinite parts, as well as an infinite body; and art, which is only the playing action of nature and a particular creature, can easier divide and separate parts than unite and make parts; for art cannot match, unite, and join parts so as nature does; for nature is not only dividable and composable, being a corporeal substance, but she is also full of curiosity and variety, being partly self-moving. And there is a great difference between forced actions and natural actions; for the one sort is [345] regular, the other irregular.

But you may say, irregularities are as natural as regularities. I grant it; but nature leaves the irregular part most commonly to her daughter or creature, art, that is, she makes irregularities for variety's sake, but she herself orders the regular part, that is, she is more careful of her regular actions; and thus nature, taking delight in variety, suffers irregularities; for otherwise, if there were only regularities, there could not be so much variety.

Again your author says that "a disease does not consist but in living bodies."[56] I answer, there is not any body that has not life; for if life is general, then all figures or parts have life; but though all bodies have life, yet all bodies have not diseases; for diseases are but accidental to bodies, and are nothing else but irregular motions in particular creatures, which be not only in animals, but generally in all creatures; for there may be irregularities in all sorts of creatures, which may cause untimely dissolutions; but yet all dissolutions are not made by irregular motions, for many creatures dissolve regularly, but only those which are untimely.

In the same place your author mentions that "a disease consists immediately in life itself, but not in the dregs and filthinesses, which are erroneous foreigners and strangers to the life."[57] I grant that a disease is made by the motions of life, but not such a life as your author describes, which does go out like the snuff of a candle, or as one of Lucian's poetical lights;[58] but by the life of nature, which cannot go out without the destruction of infinite nature. And as the motions of nature's life make diseases or irregularities, so they make that which man names "dregs" and "filths"; which dregs, filths, sickness, and [346] death are nothing but changes of corporeal motions, different from those motions or actions that are proper to the health, perfection, and consistence of such or such a figure or creation.

---

56. Van Helmont, *Oriatrike*, chapter 67, "The Subject of Inhering, of Diseases, is in the Point of Life."

57. Van Helmont, *Oriatrike*, chapter 68, "I Proceed Unto the Knowledge of Diseases."

58. In his fantastical tale *A True History*, the second-century satirist Lucian of Samosata describes the city of Lychnopolis, inhabited by lights rather than humans, which die by being snuffed out.

But, to conclude, there is no such thing as corruption, sickness, or death properly in nature, for they are made by natural actions and are only varieties in nature, but not obstructions or destructions of nature, or annihilations of particular creatures, and so is that we name "superfluities," which bear only a relation to a particular creature, which has more motion and matter than is proper for the nature of its figure. And thus much of this subject for the present, from, Madam,

Your faithful friend and humble servant.

27.

Madam,

[348] [. . .] First, concerning the cause or original of death: "Neither God," says he, "nor the evil spirit is the creator of death, but man only, who made death for himself. Neither did nature make death, but man made death natural."[59] Which if it be so, then death being, to my opinion, a natural creature, as well as life, sickness, and health, man certainly had great power as to be the creator of a natural creature. But I would fain know the reason why your author is so unwilling to make God the author of death and sickness, as well [349] as of damnation? Does it imply any impiety or irreligiousness? Does not God punish as well as reward? And is not death a punishment for our sin?

You may say, death came from sin, but sin did not come from God. Then some might ask from whence came sin? You will say, from the transgression of the command of God, as the eating of the forbidden fruit. But from whence came this transgression? It might be answered, from the persuasion of the serpent. From whence came this persuasion? From his ill and malicious nature to oppose God and ruin the race of mankind. From whence came this ill nature? From his fall. Whence came his fall? From his pride and ambition to be equal with God. From whence came this pride? From his free will. From whence came his free will? From God. Thus, Madam, if we should be too inquisitive into the actions of God, we should commit blasphemy, and make God cruel, as to be the cause of sin and consequently of damnation.

---

59. Van Helmont, *Oriatrike*, chapter 112, "The Position."

But although God is not the author of sin, yet we may not stick to say that he is the author of the punishment of sin, as an act of his divine justice, which punishment is sickness and death; nay, I see no reason why not of damnation too, as it is a due punishment for the sins of the wicked, for though man effectively works his own punishment, yet God's justice inflicts it. Like as a just judge may be called the cause of a thief being hanged.

But these questions are too curious; and some men will be as presumptuous as the devil to inquire into God's secret actions, although they be sure that they cannot be known by any creature. Wherefore let us banish such thoughts, and only [350] admire, adore, love, and praise God, and implore his mercy to give us grace to shun the punishments for our sins by the righteousness of our actions, and not endeavor to know his secret designs. [. . .]

[351] [. . .] Wherefore, leaving needless abstractions to fancies abstracted from right sense and reason, I rest, Madam,

Your faithful friend and servant.

28.

Madam,

I am very much troubled to see your author's works filled with so many spiteful reproaches and bitter taunts against the schools of physicians, condemning both their theory and practice; nay, that not only the modern schools of physicians, but also the two ancient and famous physicians, Galen and Paracelsus, must sufficiently suffer by him; especially Galen. For there is hardly a chapter in all his works which has not some accusations of blind errors, sloth, and sluggishness, ignorance, covetousness, cruelty, and the like. Which I am very sorry for; not only for the sake of your author himself, who herein does betray both his rashness and weakness in not bridling his passions, and his too great presumption, reliance, and confidence in his own abilities and extraordinary gifts; [352] but also for the sake of the fame and repute of our modern physicians; for without making now any difference between the Galenists and Paracelsians, and examining which are the best (for I think them both excellent in their kinds, especially when joined together), I will only say this in general, that the art of physic has never flourished better than now, neither has any age had more skillful, learned, and experienced physicians than this present; because they have not only the knowledge and practice of those in ages past, but also their own experience joined with it, which cannot but add perfection to their art; and I, for my part, am so much for the old way of practice that if I should be sick, I would desire rather such physicians which follow the same way, than those that by their new inventions perchance cure one and kill a hundred. [. . .]

But this I may say, that it is not always for want of skill and industry in a physician that a cure is not effected, but it lies either in the incurableness of the disease or any other [353] external accidents that do hinder the success. Not but that the best physicians may err in a disease or mistake the patient's inward distemper by his outward temper, or the interior temper by his outward distemper, or any other ways; for they may easily err through the variation of the disease, which may vary so suddenly and oft as it is impossible to apply so fast and so many medicines as the alteration requires, without certain death; for the body is not able, oftentimes, to dispose and digest several medicines so fast as the disease may vary, and therefore what was good in this temper, may, perhaps, be bad in the variation; insomuch that one medicine may in a minute prove a cordial[60] and poison.

Nay, it may be that some physicians do err through their own ignorance and mistake, must we therefore condemn all the skill, and accuse all the schools of negligence, cruelty, and ignorance? God forbid; for it would be a great injustice. Let us rather praise them for the good they do and not rashly condemn them for the evil they could not help. [. . .]

[354][. . .] But as I respect the art of physic as a singular gift from God to mankind, so I respect and esteem also learned and skillful physicians for their various knowledge, industrious studies, careful practice, and great experiences, and think everyone is bound to do the like, they being the only supporters and restorers of human life and health. For though I must confess, with your author, that God is the only giver of good, yet God is not pleased to work miracles ordinarily, but has ordained means for the restoring of health, which the art of physic does apply; and therefore those persons that are sick do wisely to send for a physician; for art, although it is but a particular creature and the handmaid of nature, yet she does nature oftentimes very good service; and so do physicians often prolong their patients' lives. The like do surgeons; for if those persons that have been wounded had been [355] left to be cured only by the magnetic medicine, I believe numbers that are alive would have been dead, and numbers would

---

60. Cordial: healing medicine, food, or drink.

die that are alive; insomuch as none would escape but by miracle, especially if dangerously hurt. [. . .]

So leaving them, I rest, Madam,

Your constant friend and faithful servant.

[356]   29.

Madam,

I am of your author's mind that "heat is not the cause of digestion"; but I dissent from him when he says that it is "the ferment of the stomach that does cause it."[61] For, in my opinion, digestion is only made by regular digestive motions, and ill digestion is caused by irregular motions, and when those motions are weak, then there is no digestion at all, but what was received remains unaltered; but when they are strong and quick, then they make a speedy digestion.

You may ask me, what are digestive motions? I answer, they are transchanging or transforming motions; but since there be many sorts of transchanging motions, digestive motions are those which transchange food into the nourishment of the body, and dispose properly, fitly, and usefully of all the parts of the food, as well as those which are converted into nourishment as of those which are cast forth.

For give me leave to tell you, Madam, that some parts of natural matter do force or cause other parts of matter to move and work according to their will, without any change or alteration of their parts; as for example, fire and metal; for fire will cause metal to flow, but it does not readily alter it from its nature of being metal; neither does fire alter its nature from being fire. And again, some parts of matter will cause other parts to work and act to their own will, by forcing these overpowered parts to [357] alter their own natural motions into the motions of the victorious party, and so transforming them wholly into their own figure; as for example, fire will cause wood to move so as to take its figure, to wit, the figure of fire, that is, to change its own figurative motions into the motions of fire. And this latter kind of moving or working is found in digestion; for the regular digestive motions do turn all food received from its own nature or figure

---

61. Van Helmont, *Oriatrike*, chapter 27, "Heat Doth Not Digest Efficiently, but Only Excitatively, or by Stirring Up."

into the nourishment, figure, or nature of the body, as into flesh, blood, bones, and the like.

But when several parts of matter meet or join with equal force and power, then their several natural motions are either quite altered or partly mixed. As for example, some received things not agreeing with the natural constitution of the body, the corporeal motions of the received and those of the receiver do dispute or oppose each other. For the motions of the received, not willing to change their nature conformable to the desire of the digestive motions, do resist, and then a war begins, whereby the body suffers most; for it causes either a sickness in the stomach, or a pain in the head, or in the heart, or in the bowels, or the like. Nay, if the received food gets an absolute victory, it dissolves and oftentimes alters the whole body, itself remaining entire and unaltered, as is evident in those that die of surfeits. But most commonly these strifes and quarrels, if violent, do alter and dissolve each other's forms or natures. [. . .]

[359][. . .] And as I cannot conceive anything that is beyond matter or a body; so I believe, according to my reason, that there is not any part in nature, be it never so subtle or small, but is a self-moving substance, or endued with self-motion; and according to the regularity and irregularity of these motions, all natural effects are produced, either perfect or imperfect; timely births or untimely and monstrous births;[62] death, health, and diseases, good and ill dispositions, natural and extravagant appetites and passions (I say natural, that is, according to the nature of their figures); sympathy and antipathy, peace and war, rational and fantastical opinions. Nevertheless, all these motions, whether regular or irregular, are natural; for regularity [360] and irregularity have but a respect to particulars and to our conceptions, because those motions which move not after the ordinary, common, or usual way or manner, we call irregular. But the curiosity and variety in nature is inconceivable by any particular; and so leaving it, I rest, Madam,

Your faithful friend and servant.

---

62. In this era, "monstrous birth" or "monster" was used to characterize any plant or animal (including humans) with a congenital defect.

34.

Madam,

Your author is not only against phlebotomy or blood-letting, but against all purging medicines, which he condemns to "carry a hidden poison in them, and to be a cruel and stupid invention."[63] [376] [. . .] But medicines are often wrong applied, and many times the disease is so various that it is as hard for a physician to hit right with several medicines as for a gunner or shooter to kill with powder and small shot a bird flying in the air; not that it is not possible to be done, but it is not ordinary or frequent. Neither does the fault lie only in the gun, powder, or shot, but in the swiftness of the flight of the bird, or in the various motion of the air, or in a sudden wind, or mist, or the like; for the same gunner may perhaps easily kill a bird sitting in a bush or hopping upon the ground. The like may be said of diseases, physicians, and medicines; for some diseases have such sudden alterations, by the sudden changes of motions, that a wise physician will not, nor cannot venture to apply so many several medicines so suddenly as the alteration requires; and shall therefore physicians be condemned? And not only condemned for what cannot be helped by reason of the variety of irregular motions, but what cannot be helped in nature? For some diseases are so deadly as no art can cure them, when as otherwise physicians with good and proper medicines have and do as yet rescue more people from death than the laws do from ruin. Nay, I have known many that have been great enemies to physic die in the flower of their age, when as others which used themselves to physic have lived a very long time.

But you may say, country-people and laborers take little or no physic, and yet grow most commonly old, whereas on the contrary, great and rich persons take much physic and do not live so long as the common sort of men does. I answer, it is [377] to be observed, first, that there are more commons than nobles, or great and rich persons; and there is not so much notice taken of the death of a mean[64] as of a noble, great, or rich person; so that for want of information or knowledge, one may easily be deceived in the number of each sort of person.

Next, the vulgar sort use laborious exercises and spare diet; when as noble and rich persons are most commonly lazy and luxurious, which breeds superfluities of humors, and these again breed many distempers. For example, you shall find few poor men troubled with the gout, stone, pox, and the like diseases,

---

63. Van Helmont, *Oriatrike*, "Of Fevers," chapter 5, "Purging Is Examined."

64. Mean: of low social status.

nor their children with rickets; for all this comes by luxury, and no doubt but all other diseases are sooner bred with luxury than temperance; but whatsoever is superfluous may, if not be taken away, yet [be] mediated with lenitive[65] and laxative medicines.

But as for physicians, surely never age knew any better, in my opinion, than this present, and yet most of them follow the rules of the schools, which are such as have been grounded upon reason, practice, and experience, for many ages. Wherefore those that will wander from the schools, and follow new and unknown ways, are, in my opinion, not orthodoxes but heretics in the art of physic. [. . .]

[378] [. . .] And it is well known to experienced physicians that medicines prepared by the art of fire are more poisonous and dangerous than natural drugs; nay, I daresay that many chemical medicines which are thought to be cordials, and have been given to patients for that purpose, have proved more poisonous than any purging physic. [. . .]

[380] [. . .] But few do consider or observe sufficiently the variety of nature's actions and the motions of particular natural creatures, which is the cause they have no better success in their cures. And so leaving them to a more diligent inquisition and search into nature and her actions, I rest, Madam,

Your faithful friend and humble servant.

[388]   37.
———————————————————————————————————

Madam,

[. . .] [390] [. . .] Your author rehearses some strange examples of child-bearing women who, having seen terrible and cruel [391] sights, as executions of malefactors and dismembering of their bodies, have brought forth monstrous births, without heads, hands, arms, legs, etc., according to the objects they had seen.[66] I must confess, Madam, that all creatures are not always formed perfect; for nature works irregularly sometimes, wherefore a child may be born defective in some member or other, or have double members instead of one, and so may other animal creatures; but this is nevertheless natural, although irregular to us.

———————————————
65. Lenitive: soothing.
66. Van Helmont, *Oriatrike*, chapter 80, "Of Material Things Injected or Cast Into the Body."

But to have a child born perfect in the womb, and the lost member to be taken off there, and so brought forth defective, as your author mentions, cannot enter my belief; neither can your author himself give any reason, but he makes only a bare relation of it; for certainly, if it was true, that the member was chopped, rent, or plucked off from the whole body of the child, it could not have been done without a violent shock or motion of the mother, which I am confident would never have been able to endure it; for such a great alteration in her body would of necessity, besides the death of the child, have caused a total dissolution of her own animal parts, by altering the natural animal motions. But, as I said above, those births are caused by irregular motions, and are not frequent and ordinary; for if upon every strange sight or cruel object, a child-bearing woman should produce such effects, monsters would be more frequent than they are. In short, nature loves variety, and this is the cause of all strange and unusual natural effects; and so leaving nature to her will and pleasure, my only delight and pleasure is to be, Madam,

[Your] faithful friend and humble servant.

[394]   **39.**

Madam,

I will not dispute your author's opinion concerning the plague of men, which he says "does not infect beasts, neither does the plague of beasts infect men,"[67] but rather believe it to be so. For I have observed that beasts infect only each other, to wit, those of their own kind, as men do infect other men. For example: the plague amongst horses continues in their own kind, and so does the plague amongst sheep; and for anything we know, there may be a plague amongst vegetables, as well as amongst animals, and they may not only infect each other, but also those animals that do feed on those infectious vegetables. So that infections may be caused several ways; either by inbreathing and [395] attracting or sucking in the poison of the plague, or by eating and converting it into the substance of the body; for some kinds of poison are so powerful as to work only by way of inbreathing.

---

67. Van Helmont, *Oriatrike*, "Tumulus Pestis, or, the Plague-Grave," chapter 17.

Also, some sorts of air may be full of infection, and infect many men, beasts, birds, vegetables, and the like; for infections are variously produced, internally as well as externally, amongst several particular creatures; for as the plague may be made internally or within the body of a particular creature, without any exterior infection entering from without into the body, so an external infection again may enter several ways into the body.

And thus there be many contagious diseases caused merely by the internal motions of the body, as by fright, terror, conceit, fancy, imagination, and the like, and many by the taking of poisonous matter from without into the body; but all are made by the natural motions of actions of animate matter, by which all is made that is in nature, and nothing is new, as Solomon says;[68] but what is thought or seems to be new is only the variation of the motions of this old matter, which is nature. And this is the reason that not every age, nation, or creature has always the like diseases; for as all the actions of nature vary, so also do diseases. [. . .]

[396] [. . .] And so I rest, Madam,

Your faithful friend and servant.

[406]   **43.**

Madam,

Your author is pleased to relate a story of one that died suddenly, and being dissected, there was not the least sign of decay or disorder found in his body.[69] But I cannot add to those that wonder when no sign of distemper is found in a man's body after he is dead; because I do not believe that the subtlest, learnedest, and most practiced anatomist can exactly tell all the interior government or motions, or can find out all obscure and invisible passages in a man's body; for concerning the motions, they are all altered in death, or rather in the dissolution of the animal figure; and although the exterior animal figure or shape does not alter so soon, yet the animal motions may alter in a moment of time; which sudden alteration may cause a sudden death, and so, the motions being invisible,

---

68. Eccles. 1:9.

69. Van Helmont, *Oriatrike*, chapter 61, "The Preface." Van Helmont describes dissecting the body of a man in order to find the cause of his death, but finding that he could observe nothing "which might be fitly accused."

the cause of death cannot be perceived; for nobody can find that which is not to be found, to wit, animal motions in a dead man; for nature has altered these motions from being animal motions to some other kind of motions, she being as various in dissolutions as in productions, indeed so various that her ways cannot be traced or known thoroughly and perfectly but only by piecemeals, as the saying is, that is, but partly. Wherefore man can only know that which is visible or subject to his senses; and yet our senses do not always inform us truly, but the alterations of grosser parts are [407] more easily known than the alterations of subtle corporeal motions, either in general or in particular; neither are the invisible passages to be known in a dead carcass, much less in a living body.

But, I pray, mistake me not when I say that the animal motions are not subject to our exterior senses; for I do not mean all exterior animal motions, nor all interior animal motions; for though you do see no interior motion in an animal body, yet you may feel some, as the motion of the heart, the motion of the pulse, the motion of the lungs, and the like; but the most part of the interior animal motions are not subject to our exterior senses; nay, no man, he may be as observing as he will, can possibly know by his exterior senses all the several and various interior motions in his own body, nor all the exterior motions of his exterior parts. And thus it remains still, that neither the subtlest motions and parts of matter, nor the obscure passages in several creatures, can be known but by several parts; for what one part is ignorant of, another part is knowing, and what one part is knowing, another part is ignorant thereof; so that unless all the parts of infinite matter were joined into one creature, there can never be in one particular creature a perfect knowledge of all things in nature. Wherefore I shall never aspire to any such knowledge, but be content with that little particular knowledge nature has been pleased to give me, the chief of which is that I know myself, and especially that I am, Madam,

Your constant friend and faithful servant.

[408] **44.**

Madam,

I perceive you are desirous to know the cause, "Why a man is more weak at the latter end of a disease than at the beginning, and is a longer time recovering health, than losing health; as also the reason of relapses and intermissions?"

First, as for weakness and strength, my opinion is, they are caused by the regular and irregular motions in several parts, each striving to overpower the other in their conflict; and when a man recovers from a disease, although the regular motions have conquered the irregular and subdued them to their obedience, yet they are not so quite obedient as they ought, which causes weakness. Neither do the regular motions use so much force in peace as in war; for though animate matter cannot lose force, yet it does not always use force; neither can the parts of nature act beyond their natural power, but they do act within their natural power; neither do they commonly act to the utmost of their power.

And as for health, why it is sooner lost than recovered, I answer that it is easier to make disorders than to rectify them. As for example, in a commonwealth, the ruins of war are not so suddenly repaired as made. But concerning relapses and intermissions of diseases, intermissions are like truces or cessations from war for a time; and relapses are like new stirs or tumults of rebellion; for rebels are not so apt to settle in [409] peace as to renew the war upon slight occasions; and if the regular motions of the body be stronger, they reduce them again unto obedience.

But diseases are occasioned many ways; for some are made by a home rebellion, and others by foreign enemies, and some by natural and eager dissolutions, and their cures are as different; but the chief magistrates or governors of the animal body, which are the regular motions of the parts of the body, want most commonly the assistance of foreign parts, which are medicines, diets, and the like; and if there be factions amongst these chief magistrates or motions of the parts of the body, then the whole body suffers ruin.

But since there would be no variety in nature, nor no difference between nature's several parts or creatures, if her actions were never different but always agreeing and constant, a war or rebellion in nature cannot be avoided. But, mistake me not, for I do not mean a war or rebellion in the nature or substance of matter, but between the several parts of matter, which are the several creatures, and their several motions; for matter being always one and the same in its nature, has nothing to war withal; and surely it will not quarrel with its own nature.

Next you desire to know, that if nature be in a perpetual motion, "whence comes a duration of some things, and a tiredness, weariness, sluggishness, or faintness?" I answer, first, that in some bodies, the retentive motions are stronger than the dissolving motions; as for example, gold, and quicksilver or mercury; the separating and dissolving motions of fire have only power to melt and rarify them for a time, but cannot alter their natures; so a hammer [410] or such like instrument, when used, may beat gold and make it thin as a cobweb or as dust, but cannot alter its interior nature. But yet this does not prove it to be either without motion or to be altogether unalterable, and not subject to any dissolution; but only that its retentive motions are too strong for the dissolving motions of the fire, which by force work upon the gold; and we might as well say that sand, or an earthen vessel, or glass, or stone, or anything else, is unalterable and will last eternally if not disturbed. But some of nature's actions are as industrious to keep their figures as others are to dissolve or alter them; and therefore retentive motions are more strong and active in some figures than dissolving motions are in others, or producing motions in other figures.

Next, as for tiredness or faintness of motions, there is no such thing as tiredness or faintness in nature, for nature cannot be tired, nor grow faint, or sick, nor be pained, nor die, nor be any ways defective; for all this is only caused through the change and variety of the corporeal motions of nature and her several parts; neither do irregular motions prove any defect in nature, but a prudence in nature's actions, in making varieties and alterations of figures; for without such motions or actions, there could not be such varieties and alterations in nature as there are. Neither is slackness of some motions a defect, for nature is too wise to use her utmost force in her ordinary works; and though nature is infinite, yet it is not necessary she should use an infinite force and power in any particular act.

Lastly, you desire my opinion, "whether there be motion in a dead animal creature." To which I answer, I have [411] declared heretofore that there is no such thing as death in nature, but what is commonly named "death" is but an alteration or change of corporeal motions, and the death of an animal is nothing else but the dissolving motions of its figure; for when a man is dying, the motions which did formerly work to the consistence of his figure do now work to the dissolution of his figure, and to the production of some other figures, changing and transforming every part thereof; but though the figure of that dead animal is dissolved, yet the parts of that dissolved figure remain still in nature although they be infinitely changed, and will do so eternally, as long as nature lasts by the will of God; for nothing can be lost or annihilated in nature.

And this is all, Madam, that I can answer to your questions, wherein I hope I have obeyed your commands, according to the duty of, Madam,

Your faithful friend and humble servant.

45.

Madam,

I have thus far discharged my duty, that according to your commands, I have given you my judgment of the works of those four famous philosophers of our age which you did send me to peruse, and have [412] withal made reflections upon some of their opinions in natural philosophy, especially those wherein I did find them dissent from the ground and principles of my own philosophy.

And since by your leave I am now publishing all those letters which I have hitherto written to you concerning those aforesaid authors and their works, I am confident I shall not escape the censures of their followers. But I shall desire them that they will be pleased to do me this justice and to examine first my opinions well, without any partiality or willful misinterpretation of my sense, before they pass their censure.

Next, I desire them to consider that I have no skill in school-learning, and therefore for want of terms of art may easily chance to slip, or at least not express my opinions so clearly as my readers expected. However, I have done my endeavor, and to my sense and reason they seem clear and plain enough, especially as I have expressed them in those letters I have sent you; for concerning my other work, called *Philosophical Opinions*, I must confess, that it might have been done more exactly and perspicuously, had I been better skilled in such words and expressions as are usual in the schools of philosophers; and therefore, if I be but capable to learn names and terms of art (although I find myself very untoward to learn, and do despair of proving a scholar), I will yet endeavor to rectify that work and make it more intelligible; for my greatest ambition is to express my conceptions so that my readers may understand them. For which I would not spare any labor or pains, but be as industrious as those that gain their living by their work; and I pray to God that nature may give me a [413] capacity to do it.

But as for those that will censure my works out of spite and malice, rather than according to justice, let them do their worst; for if God do but bless them, I need not to fear the power of nature, much less of a part of nature as man. Nay, if I have but your Ladyship's approbation, it will satisfy me; for I know you are so wise and just in your judgment that I may safely rely upon it. For which I shall constantly and unfeignedly remain as long as I live, Madam,

Your Ladyship's most faithful friend and humble servant.

SECTION 4

1.

Madam,

I perceive you take great delight in the study of natural philosophy, since you have not only sent me some authors to peruse and give my judgment of their opinions, but are very studious yourself in the reading of philosophical works. And truly, I think you cannot spend your time more honorably, profitably, and delightfully, than in the study of nature, as to consider how variously, curiously, and wisely she acts in her creatures, for if the particular knowledge of a man's self be commendable, much more is the knowledge of the general actions of nature, which does lead us to the knowledge of our selves. The truth is, by the help of philosophy our minds are raised above ourselves into the knowledge of the causes of all natural effects.

But [415] leaving the commending of this noble study, you are pleased to desire my opinion of a very difficult and intricate argument in natural philosophy, to wit, of generation[1] or natural production. I must beg leave to tell you, first, that some (though foolishly) believe it is not fit for women to argue upon so subtle a mystery; next, there have been so many learned and experienced philosophers, physicians, and anatomists which have treated of this subject, that it might be thought a great presumption for me to argue with them, having neither the learning nor experience by practice which they had; lastly, there are so many several ways and manners of productions in nature as it is impossible for a single creature to know them all. For there are infinite variations made by self-motion in infinite matter, producing several figures, which are several creatures in that same matter.

But you would fain know, how nature, which is infinite matter, acts by self-motion? Truly, Madam, you may as well ask any one part of your body

---

1. Generation: reproduction.

how every other part of your body acts, as to ask me, who am but a small part of infinite matter, how nature works. But yet, I cannot say that nature is so obscure as her creatures are utterly ignorant; for as there are two of the outward sensitive organs in animal bodies which are more intelligible than the rest, to wit, the ear and the eye; so in infinite matter, which is the body of nature, there are two parts which are more understanding or knowing than the rest, to wit, the rational and sensitive part of infinite matter. For though it be true that nature, by self-division, made by self-motion into self-figures, which are self-parts, causes a self-obscurity to [416] each part, motion, and figure; nevertheless, nature being infinitely wise and knowing, its infinite natural wisdom and knowledge is divided amongst those infinite parts of the infinite body; and the two most intelligible parts, as I said, are the sensitive and rational parts in nature, which are divided, being infinite, into every figure or creature; I cannot say equally divided, no more than I can say all creatures are of equal shapes, sizes, properties, strengths, quantities, qualities, constitutions, semblances, appetites, passions, capacities, forms, natures, and the like; for nature delights in variety, as human sense and reason may well perceive; for seldom any two creatures are just alike, although of one kind or sort, but every creature does vary more or less.

Wherefore it is not probable that the production or generation of all or most creatures should be after one and the same manner or way, for else all creatures would be just alike without any difference. But this is to be observed, that though nature delights in variety, yet she does not delight in confusion, but as it is the propriety of nature to work variously, so she works also wisely; which is the reason that the rational and sensitive parts of nature, which are the designing and architectonical parts, keep the species of every kind of creature by the way of translation in generation or natural production; for whatsoever is transferred works according to the nature of that figure or figures from whence it was transferred. But mistake me not; for I do not mean always according to their exterior figure, but according to their interior nature; for different motions in one and the same parts of matter make different [417] figures, wherefore much more in several parts of matter and changes of motion. But, as I said, translation is the chief means to keep or maintain the species of every kind of creatures, which translation in natural production or generation is of the purest and subtlest substances, to wit, the sensitive and rational, which are the designing and architectonical parts of nature.

You may ask me, Madam, what this wise and ingenious matter is. I answer, it is so pure, subtle, and self-active, as our human shares of sense and reason cannot readily or perfectly perceive it; for by that little part of knowledge that a human creature has, it may more readily perceive the strong action than the purer substance; for the strongest action of the purest substance is more perceivable than the matter or substance itself; which is the cause that most men are apt to believe the motion and to deny the matter, by reason of its subtlety; for surely the sensitive and rational matter is so pure and subtle as not to be expressed by human sense and reason.

As for the rational matter, it is so pure, fine, and subtle that it may be as far beyond lucent matter as lucent matter is beyond gross vapors or thick clouds; and the sensitive matter seems not much less pure. Also there is very pure inanimate matter, but not subtle and active of itself; for as there are degrees in the animate, so there are also degrees in the inanimate matter; so that the purest degree of inanimate matter comes next to the animate, not in motion but in the purity of its own degree; for it cannot change its nature so as to become animate, yet it may be so pure in its own nature as not to be perceptible by our grosser sense.

[418] But concerning the two degrees of animate matter, to wit, the sensitive and rational, I say that the sensitive is much more acute than vitriol, aqua-fortis, fire, or the like;[2] and the rational much more subtle and active than quicksilver[3] or light, so as I cannot find a comparison fit to express them, only that this sensitive and rational self-moving matter is the life and soul of nature. But by reason this matter is not subject to our gross senses, although our senses are subject to it, as being made, subsisting, and acting through the power of its actions, we are not apt to believe it, no more than a simple country-wench will believe that air is a substance, if she neither hear, see, smell, taste, or touch it, although air touches and surrounds her. But yet the effects of this animate matter prove that there is such a matter; only, as I said before, this self-moving matter, causing a self-division as well as a general action, is the cause of a self-obscurity, which obscurity causes doubts, disputes, and inconstancies in human opinions, although not so much obscurity as to make all creatures blindfold, for surely there is no creature but perceives more or less.

But to conclude, the rational degree of matter is the most intelligible and the wisest part of nature, and the sensitive is the most laborious and provident part in nature, both which are the creators of all creatures in infinite matter; and if you intend to know more of this rational and sensitive matter, you may consult

---

2. Vitriol and aqua-fortis were corrosive substances used by alchemists to dissolve metals.
3. Quicksilver: mercury.

my book of philosophy,⁴ to which I refer you. And so taking my leave for the present, I rest, Madam,

Your faithful friend and servant.

[419]   **2.**
_____

Madam,

I understand by your last that you have read the book of that most learned and famous physician and anatomist, Dr. Harvey, which treats of generation;⁵ and in the reading of it, you have marked several scruples, which you have framed into several questions concerning that subject, to which you desire my answer. Truly, Madam, I am loath to embark myself in this difficult argument, not only for the reasons I have given you heretofore, but also that I do not find myself able enough to give you such a satisfactory answer as perhaps you do expect. But since your commands are so powerful with me that I can hardly resist them, and your nature so good that you easily pardon anything that is amiss, I will venture upon it according to the strength of my natural reason, and endeavor to give you my opinion as well and as clearly as I can.

Your first question is, "Whether the action of one or more producers be the only cause of natural production or generation, without imparting or transferring any of their substance or matter." I answer, the sole co-action of the producers may make a change of exterior forms or figures, but not produce another creature; for if there were not substance or matter as well as action, both transferred together, there would not be new creatures made out of old matter, but every production would require new matter, which is impossible if [420] there be but one matter, and that infinite; and certainly, human sense and reason may well perceive that there can be but one matter, for several kinds of matter would make a confusion; and thus if new creatures were made only by substanceless motion, it would not only be an infinite trouble to nature to create something out of nothing perpetually, but, as I said, it would make a confusion amongst all nature's works, which are her several parts or creatures.

_____

4. That is, *Philosophical and Physical Opinions* (1663).

5. William Harvey (1578–1657). Cavendish discusses his *Anatomical Exercitations Concerning the Generation of Living Creatures* (1653). See the Introduction (pp. xli–xlii).

But by reason there is but one matter, which is infinite and eternal, and this matter has self-motion in it, both matter and motion must of necessity transmigrate, or be transferred together without any separation, as being but one thing, to wit, corporeal motion. 'Tis true, one part of animate or self-moving matter may without translation move, or rather occasion other parts to move; but one creature cannot naturally produce another without the transferring of its corporeal motions.

But it is well to be observed that there is great difference between the actions of nature; for all actions are not generating, but some are patterning, and some transforming, and the like; and as for the transforming action, that may be without translation, as being nothing else but a change of motions in one and the same part or parts of matter, to wit, when the same parts of matter do change into several figures, and return into the same figures again. Also, the action of patterning is without translation; for to pattern out is nothing else but to imitate and to make a figure in its own substance or parts of matter like another figure. But in generation every producer does transfer both matter and motion, that is, corporeal motion, [421] into the produced; and if there be more producers than one, they all do contribute to the produced; and if one creature produces many creatures, those many creatures do partake more or less of their producer.

But you may say, if the producer transfers its own matter, or rather its own corporeal motions into the produced, many productions will soon dissolve the producer, and he will become a sacrifice to his offspring. I answer, that does not follow. For as one or more creatures contribute to one or more other creatures, so other creatures do contribute to them, although not after one and the same manner or way, but after diverse manners or ways; but all manners and ways must be by translation to repair and assist; for no creature can subsist alone and of itself, but all creatures traffic and commerce from and to each other, and must of necessity do so, since they are all parts of the same matter. Neither can motion subsist without matter, no more than an artificer can work without materials, and without self-motion, matter would be dead and useless. Wherefore matter and motion must upon necessity not only be inseparable, but be one body, to wit, corporeal motion; which motion by dividing and composing its several parts and acting variously is the cause of all production, generation, metamorphosing, or any other thing that is done in nature. But if, according to your author, the sole action be the cause of generation without transferring of substance, then matter is useless and of none or little effect; which, in my opinion, is not probable.

Your second question is, "whether the production or [422] generation of animals is as the conceptions of the brain, which the learned say are immaterial?"[6]
I answer, the conceptions of the brain, in my opinion, are not immaterial, but
corporeal; for though the corporeal motions of the brain, or the matter of its
conceptions, is invisible to human creatures, and that when the brain is dissected there is no such matter found, yet that does not prove that there is no
matter, because it is not so gross a substance as to be perceptible by our exterior
senses. Neither will your author's example hold, that as a builder erects a house
according to his conception in the brain, the same happens in all other natural
productions or generations; for in my opinion, the house is materially made in
the brain, which is the conception of the builder, although not of such gross
materials as stone, brick, wood, and the like, yet of such matter as is the rational
matter; that is, the house when it is conceived in the brain is made by the rational
corporeal figurative motions of their own substance or degree of matter. [. . .]

[423] Your fourth question is, "whether an animal creature is perfectly
shaped or formed at the first conception?" I answer, if the creature is composed
of many and different parts, it cannot be. You may say that if it has not all his
parts produced at once, there will be required many acts of generation to beget
or produce every part, otherwise the producers would not be the parents of the
produced in whole, but in part. I answer, the producer is the designer, architect,
and founder of the whole creature produced; [424] for the sensitive and rational
corporeal motions, which are transferred from the producer or producers, join
to build the produced like to the producer in specie,[7] but the transferred parts
may be invisible and insensible to human creatures, both through their purity
and little quantity, until the produced is framed to some sensible degree; for a
stately building may proceed from a small beginning, neither can human sense
tell what manner of building is designed at the first foundation. [. . .]

Again, you may say that some parts of matter may produce another creature
not like to the producer in its specie, as for example, monsters. I answer, that is
possible to be done, but yet it is not [425] usual; for monsters are not commonly
born, but those corporeal motions which dwell in one species work according to
the nature of the same species [. . .].

[426] [. . .] And to conclude this question, we may observe that not any animal creature's shape dissolves in one instant of time, but by degrees; why should
we believe, then, that animals are generated or produced in their perfect

---

6. Harvey makes this comparison in the final chapter of *Anatomical Exercitations*.

7. In specie: in the form of its species.

shape in one instant of time and by one act of nature? But sense and reason knows by observation that an animal creature requires more time to be generated than to be dissolved, like as a house is sooner and with less pains pulled down than built up.

Your fifth question is, "whether animals are not generated by the way of metamorphosing?" To which I answer that it is not possible that a third creature can be made without translation of corporeal motions; and since metamorphosing is only a change of motions in the same parts of matter, without any translation of corporeal motions, no animal creature can be produced or generated by the way of metamorphosing.

Your sixth question is, "whether a whole may be made out of a part?" I answer, there is no whole in nature, except you will call nature herself a whole; for all creatures are but parts of infinite matter.

[427] Your seventh question is, "whether all animals, as also vegetables, are made or generated by the way of eggs?" I have said heretofore that it is not probable that different sorts, nay, different kinds of creatures, should all have but one manner or way of production; for why should not nature make different ways of productions as well as different creatures? And as for vegetables, if all their seeds be likened unto eggs, then eggs may very well be likened to seeds; which if so, then a peas-cod[8] is in the hen, and the peas in the cod is the cluster of eggs; the like of ears of corn. And those animals that produce but one creature or seed at a time may be like the kernel of a nut; when the shell is broken, the creature comes forth. But how this will agree with your author, who says that the creature in the shell must make its own passage, I cannot tell; for if the nut be not broken by some external means or occasion, the kernel is not like to get forth.

And as for human eggs, I know not what to answer, for it is said that the first woman was made of a man's rib, but whether that rib was an egg, I cannot tell. And why may not minerals and elements be produced by the way of eggs as well as vegetables and animals? Nay, why may not the whole world be likened unto an egg? [. . .] [428] Or it might be said that the chaos was an egg, and the universe, the chicken.

But leaving this similizing, it is like, that some studious men may by long study upon one part of the body conceive and believe that all other parts are like that one part; like as those that have gazed long upon the sun, all they see for a time are suns to them; or like as those which having heard much of hobgoblins, all they see are hobgoblins, their fancies making such things. [. . .]

---

8. Peas-cod: peapod

Your eighth question is, "whether it may not be that the sensitive and rational corporeal motions in an egg do pattern out the figure of the hen and cock,
whilst the hen sits upon the egg, and so bring forth chickens by the way of patterning?" I answer, the action of patterning is not the action of generation; for as
I said heretofore, the actions of nature are different, and generation must needs
be performed by the way of translation, which translation is not required in the
action of patterning; but according as the producers are, which transfer their
own matter into the produced, so is the produced concerning its species, which
is plainly proved by common examples; for if pheasants, or turkey, or goose eggs
be laid under an ordinary hen, or an ordinary hen's egg be laid under a pheasant,
turkey, or goose, the chickens of those eggs will never be of any other species
than of those [429] that produced the egg; for an ordinary hen, if she sit upon
pheasants, turkey, or goose-eggs, does not hatch chickens of her own species,
but the chickens will be of the species either of the pheasant, or turkey, or goose
which did at first produce the egg; which proves that in generation, or natural
production, there is not only required the action of the producers, but also a
transferring of some of their own parts to form the produced. But you may say,
what does the sitting hen contribute then to the production of the chicken? I
answer, the sitting hen does only assist the egg in the production of the chicken,
as the ground does the seed.

Your ninth question is, "concerning the soul of a particular animal creature,
as whether it be wholly of itself, and subsist wholly in and by itself?" But you
must give me leave first to ask you what soul you mean, whether the divine or
the natural soul, for there is great difference between them, although not the
least that ever I heard rightly examined and distinguished; and if you mean the
divine soul, I shall desire you to excuse me, for that belongs to divines and not to
natural philosophers; neither am I so presumptuous as to entrench upon their
sacred order. But as for the natural soul, the learned have divided it into three
parts, to wit, the vegetative, sensitive, and rational soul; and according to these
three souls, made three kinds of lives, as the vegetative, sensitive, and rational
life.[9] But they might as well say there are infinite bodies, lives, and souls, as three;
for in nature there is but one life, soul, and body, consisting all of one matter,
which is corporeal nature. But yet by reason this life [430] and soul is material,
it is divided into numerous parts, which make numerous lives and souls in every
particular creature; for each particular part of the rational self-moving matter

---

9. In *De Anima*, book 2, Aristotle distinguishes between the vegetative or nutritive soul (possessed by plants, animals, and humans), the sensitive soul (possessed by animals and humans), and
the rational soul (possessed only by humans).

is each particular soul in each particular creature, but all those parts considered in general make but one soul of nature; and as this self-moving rational matter has power to unite its parts, so it has ability or power to divide its united parts.

And thus the rational soul of every particular creature is composed of parts (I mean parts of a material substance; for whatsoever is substanceless and incorporeal belongs not to nature, but is supernatural), for by reason the infinite and only matter is by self-motion divided into self-parts, not any creature can have a soul without parts; neither can the souls of creatures subsist without commerce of other rational parts, no more than one body can subsist without the assistance of other bodies; for all parts belong to one body, which is nature; nay, if anything could subsist of itself, it were a god and not a creature. Wherefore not any creature can challenge a soul absolutely to himself, unless man, who has a divine soul, which no other creature has.

But that which makes so many confusions and disputes amongst learned men is that they conceive, first, there is no rational soul but only in man; next, that this rational soul in every man is individable. But if the rational soul is material, as certainly to all sense and reason it is, then it must not only be in all material creatures, but be dividable too; for all that is material or corporeal has parts and is dividable, and therefore there is no such thing in any one creature as one entire soul; nay, we might as well say [431] there is but one creature in nature, as say there is but one individable natural soul in one creature.

Your tenth question is, "whether souls are producible, or can be produced?" I answer, in my opinion, they are producible, by reason all parts in nature are so. But mistake me not, for I do not mean that any one part is produced out of nothing, or out of new matter; but one creature is produced by another, by the dividing and uniting, joining and disjoining of the several parts of matter, and not by substanceless motion out of new matter. And because there is not anything in nature that has an absolute subsistence of itself, each creature is a producer as well as a produced, in some kind or other; for no part of nature can subsist single and without reference and assistance of each other, or else every single part would not only be a whole of itself, but be as a god without control; and though one part is not another part, yet one part belongs to another part, and all parts to one whole, and that whole to all the parts, which whole is one corporeal nature. [. . .]

[432] [. . .] I'll add no more, but repeat what I said in the beginning, *viz.*, that I rely upon the goodness of your nature, from which I hope for pardon if I have not so exactly and solidly answered your desire; for the argument of this discourse, being so difficult, may easily lead me into an error, which your better

judgment will soon correct; and in so doing you will add to those favors for which I am already, Madam,

Your Ladyship's most obliged friend and humble servant.

[433]　　　　　　　　　　　**3.**

Madam,

You thought verily I had mistaken myself in my last, concerning the rational souls of every particular creature, because I said all creatures had numerous souls; and not only so, but every particular creature had numerous souls. Truly, Madam, I did not mistake myself, for I am of the same opinion still; for though there is but one soul in infinite nature, yet that soul being dividable into parts, every part is a soul in every single creature, were the parts no bigger in quantity than an atom.

But you ask whether nature has infinite souls? I answer that infinite nature is but one infinite body, divided into infinite parts, which we call "creatures"; and therefore it may as well be said that nature is composed of infinite creatures or parts as [that] she is divided into infinite creatures or parts; for nature, being material, is dividable and composable. The same may be said of nature's soul, which is the rational part of the only infinite matter, as also of nature's life, which is the sensitive part of the only infinite self-moving matter; and of the inanimate part of the only infinite matter, which I call the "body" for distinction's sake, as having no self-motion in its own nature, for infinite material nature has an infinite material soul, life, and body.

But, Madam, I desire you to observe what I said already, *viz.*, [434] that the parts of nature are as apt to divide as to unite; for the chief actions of nature are to divide and to unite; which division is the cause that it may well be said, every particular creature has numerous souls; for every part of rational matter is a particular soul, and every part of the sensitive matter is a particular life; all which, mixed with the inanimate matter, though they be infinite in parts, yet they make but one infinite whole, which is infinite nature; and thus the infinite division into infinite parts is the cause that every particular creature has numerous souls, and the transmigration of parts from and to parts is the reason that not any creature can challenge a single soul or souls to itself; the same for life [. . .]

But to conclude, those creatures which have their rational parts most united are the wisest; and those that have their rational parts most divided are the wittiest; and those that have much of this rational matter are much knowing; and those which have less of this rational matter are less knowing; and there is no creature that has not some; for like as all the parts of a human body are endued with life and soul, so are all the parts of infinite nature; [435] and though some parts of matter are not animate in themselves, yet there is no part that is not mixed with the animate matter; so that all parts of nature are moving and moved. And thus, hoping I have cleared myself in this point to your better understanding, I take my leave and rest, Madam,

Your faithful friend and servant.

4.

Madam,

In the works of that most famous philosopher and mathematician of our age, Gal.,[10] which you thought worth my reading, I find he discourses much of upward and downward, backward and forward; but to tell you really, I do not understand what he means by those words, for in my opinion, there is properly no such thing as upward, downward, backward, or forward in nature, for all this is nothing else but natural corporeal motions, to which in respect of some particulars we do attribute such or such names; for if we conceive a circle, I pray where is upward and downward, backward and forward? Certainly it is, in my opinion, just like that they name "rest," "place," "space," "time," etc., when as nature herself knows of no such things, but all these are only [436] the several and various motions of the only matter.

You will say, how can rest be a motion? I answer, "rest" is a word which expresses rather man's ignorance than his knowledge; for when he sees that a particular creature has not any external local motion perceptible by his sight, he says it rests, and this rest he calls a cessation from motion, when as yet there is no such thing as cessation from motion in nature; for motion is the action of

---

10. Galileo Galilei (1564–1642). Although Cavendish does not say so, the book she is discussing is Galileo's *Dialogue Concerning the Two Chief World Systems* (1632). See the Introduction (p. xliii).

natural matter, and its nature is to move perpetually; so that it is more probable for motion to be annihilated than to cease. But you may say it is a cessation from some particular motion. I answer, you may rather call it an alteration of a particular motion than a cessation; for though a particular motion does not move in that same manner as it did before, nevertheless it is still there, and not only there, but still moving; only it is not moving after the same manner as it did move heretofore, but has changed from such a kind of motion to another kind of motion, and, being still moving, it cannot be said to cease. Wherefore what is commonly called cessation from motion is only a change of some particular motion, and is a mistake of change for rest.

Next, I find in the same author a long discourse of circular and straight motion, to wit, "that they are simple motions, and that all others are composed out of them, and are mixed motions; also, that the circular motion is perfect, and the right imperfect; and that all the parts of the world, if moveable of their own nature, it is impossible that their motions should be right, or any other than circular; that a circular motion is never to be gotten naturally, without a preceding right motion; that a right motion cannot [437] naturally be perpetual; that a right motion is impossible in the world well ordered; and the like."[11] First, I cannot conceive why natural matter should use the circle-figure more than any other in the motions of her creatures; for nature, which is infinite matter, is not bound to one particular motion, or to move in a circle more than any other figure, but she moves more variously than any one part of hers can conceive. Wherefore it is not requisite that the natural motions of natural bodies should be only circular.

Next, I do not understand why a circular motion cannot be gotten naturally without a precedent right motion; for in my opinion, corporeal motions may be round or circular, without being or moving straight before; and if a straight line does make a circle, then an imperfect figure makes a perfect; but in my opinion, a circle may as well make a straight line as a straight line a circle; except it be like a Gordian knot,[12] that it cannot be dissolved, or that nature may make some corporeal motions as constant as she makes others inconstant, for her motions are not alike in continuance and alteration.

---

11. Cavendish summarizes various claims about motion endorsed by the character Salviati (representing Galileo's views) in the first dialogue of Galileo's *Dialogue*, "The First Day."

12. According to legend, Alexander the Great fulfilled a prophecy that whoever could untangle a mass of knots tying up Gordius's wagon would become ruler of Asia. Alexander solved the problem by slicing through the knots with his sword.

And as for right motion, that naturally it cannot be perpetual; my opinion is that it cannot be, if nature be finite; but if nature be infinite, it may be. But the circular motion is more proper for a finite than an infinite, because a circle-figure is perfect and circumscribed, and a straight line is infinite, or at least producible in infinite; and there may be other worlds in infinite nature besides these round globes perceptible by our sight, which may have other figures; for though it be proper for globes or spherical bodies to move round, yet that does not prove that infinite matter moves [438] round, or that all worlds must be of a globous figure; for there may be as different worlds as other creatures.

He says that a right motion is impossible in the world well ordered; but I cannot conceive a right motion to be less orderly than a circular in nature, except it be in some particulars; but oftentimes that which is well ordered in some cases seems to some men's understandings and perceptions ill ordered in other cases; for man, as a part, most commonly considers but the particulars, not the generals, like as everyone in a commonwealth considers more himself and his family than the public.

Lastly, concerning the simplicity of motions, as that only circular and straight motions are simple motions, because they are made by simple lines; I know not what they mean by simple lines; for the same lines which make straight and circular figures may make as well other figures as those; but in my opinion, all motions may be called simple in regard of their own nature; for they are nothing else but the sensitive and rational part of matter, which in its own nature is pure and simple, and moves according to the nature of each figure, either swiftly or slowly, or in this or that sort of motion; but the most simple, purest, and subtlest part is the rational part of matter, which though it be mixed with the sensitive and inanimate in one body, yet it can and does move figuratively in its own matter, without the help or assistance of any other.

But I desire you to remember, Madam, that in the compositions and divisions of the parts of nature there is as much unity and agreement as there is discord and disagreement; for in infinite, there is no such thing as most and least; neither is there [439] any such thing as more perfect or less perfect in matter. And as for irregularities, properly there is none in nature, for nature is regular; but that which man (who is but a small part of nature, and therefore but partly knowing) names "irregularities" or "imperfections" is only a change and alteration of motions; for a part can know the variety of motions in nature no more than finite can know infinite, or the bare exterior shape and figure of a man's body can know the whole body, or the head can know the mind; for infinite natural knowledge is corporeal; and being corporeal, it is dividable; and being dividable, it cannot be confined to one part only; for there is no such thing as an absolute determination or subsistence in parts without relation or dependence upon one another.

And since matter is infinite, and acts wisely, and all for the best, it may be as well for the best of nature when parts are divided antipathetically as when they are united sympathetically. Also matter being infinite, it cannot be perfect, neither can a part be called perfect, as being a part. But mistake me not, Madam; for when I say there is no perfection in nature, as I do in my *Philosophical Opinions*,[13] I mean by perfection a finiteness, absoluteness, or completeness of figure; and in this sense I say nature has no perfection, by reason it is infinite; but yet I do not deny but that there is a perfection in the nature or essence of infinite matter; for matter is perfect matter; that is, pure and simple in its own substance or nature, as mere matter, without any mixture or addition of some thing that is not matter, or that is [440] between matter and no matter; and material motions are perfect motions, although infinite; just as a line may be called a perfect line, although it be endless, and gold or any metal may be called perfect gold or perfect metal, though it be but a part. And thus it may be said of infinite nature, or infinite matter, without any contradiction, that it is both perfect and not perfect; perfect in its nature or substance, not perfect in its exterior figure.

But you may say, if infinite matter be not perfect, it is imperfect, and what is imperfect wants something. I answer, that does not follow; for we cannot say that what is not perfect must of necessity be imperfect, because there is something else which it may be, to wit, infinite; for as imperfection is beneath perfection, so perfection is beneath infinite; and though infinite matter be not perfect in its figure, yet it is not imperfect but infinite; for perfection and imperfection belong only to particulars and not to infinite. And thus much for the present. I conclude, and rest Madam,

Your Ladyship's most obliged friend and humble servant.

---

13. *Philosophical and Physical Opinions* (1663), part 1, chapter 14.

[441]   **5.**

Madam,

The author mentioned in my former letter says "that quietness is the degree of infinite slowness, and that a moveable body passing from quietness, passes through all the degrees of slowness without staying in any."[14] But I cannot conceive that all the parts of matter should be necessitated to move by degrees; for though there be degrees in nature, yet nature does not in all her actions move by degrees. You may say, for example, from one to twenty, there are eighteen degrees between one and twenty; and all these degrees are included in the last degree, which is twenty. I answer, that may be. But yet there is no progress made through all those degrees; for when a body does move strong at one time, and the next after moves weak, I cannot conceive how any degrees should really be made between.

You may say, by imagination. But this imagination of degrees is like the conception of space and place, when as yet there is no such thing as place or space by itself; for all is but one body, and motion is the action of this same body, which is corporeal nature; and because a particular body can and does move after various manners, according to the change of its corporeal motions, this variety of motions man calls "place," "space," "time," "degrees," etc., considering them by themselves and giving them peculiar names, as if they could [442] be parted from body, or at least be conceived without body; for the conception or imagination itself is corporeal, and so are they nothing else but corporeal motions.

But it seems as if this same author conceived also motion to be a thing by itself, and that motion begets motion, when he says that a body by moving grows stronger in motion by degrees, when as yet the strength was in the matter of the body eternally; for nature was always a grave matron, never a suckling infant. And though parts by dissolving and composing may lose and get acquaintance of each other, yet no part can be otherwise in its nature than it ever was. Wherefore change of corporeal motions is not losing nor getting strength or swiftness; for nature does not lose force, although she does not use force in all her various actions; neither can any natural body get more strength than by nature it has, although it may get the assistance of other bodies joined to it.

But swiftness and slowness are according to the several figurative actions of self-moving matter; which several actions or motions of nature, and their

---

14. Galileo, *Dialogue*, "The First Day." The word translated as "quietness" in the version Cavendish was using can also be translated as "rest."

alterations, cannot be found out by any particular creature. As for example, the motions of lead and the motions of wood, unless man know their several causes; for wood, in some cases, may move slower than lead; and lead, in other cases, slower than wood.

Again, the same author says "that a heavy moveable body descending gets force enough to bring it back again to as much height."[15] But I think it might as well be said that a man walking a mile gets as much strength as to walk back that mile; when 'tis likely that having walked ten miles, he may not have so much strength as to walk back again one [443] mile; neither is he necessitated to walk back, except some other more powerful body do force him back. For though nature is self-moving, yet every part has not an absolute power, for many parts may overpower fewer; also several corporeal motions may cross and oppose as well as assist each other; for if there were not opposition, as well as agreement and assistance amongst nature's parts, there would not be such variety in nature as there is.

Moreover, he makes mention of a "line, with a weight hung to its end, which being removed from the perpendicular, presently falls to the same again."[16] To which I answer, that it is the appetite and desire of the line not to move by constraint or any forced exterior motion; but that which forces the line to move from the perpendicular does not give it motion, but is only an occasion that it moves in such a way; neither does the line get that motion from any other exterior body, but it is the line's own motion; for if the motion of the hand or any other exterior body should give the line that motion, I pray, from which does it receive the motion to tend to its former state? Wherefore, when the line moves backward or forward, it is not that the line gets what it had not before, that is, a new corporeal motion, but it uses its own motion; only, as I said, that exterior body is the occasion that it moves after such a manner or way, and therefore this motion of the line, although it is the line's own motion, yet in respect of the exterior body that causes it to move that way, it may be called a "forced," or rather an "occasioned" motion.

And thus no body can get motion from another body, except it get matter too; for all that motion that a body has proceeds from the self-moving part of matter, and [444] motion and matter are but one thing; neither is there any inanimate part of matter in nature which is not commixed with the animate, and consequently, there is no part which is not moving or moved; the animate

---

15. Galileo, *Dialogue*, "The First Day."
16. Galileo, *Dialogue*, "The First Day."

part of matter is the only self-moving part, and the inanimate the moved. Not that the animate matter does give away its own motion to the inanimate, and that the inanimate becomes self-moving; but the animate, by reason of the close conjunction and commixture, works together with the inanimate, or causes the inanimate to work with it; and thus the inanimate remains as simple in its own nature as the animate does in its nature, although they are mixed; for those mixtures do not alter the simplicity of each other's nature. But having discoursed of this subject in my former letters, I take my leave, and rest, Madam,

Your faithful friend and servant.

6.

Madam,

It seems my former letter concerning motion has given you occasion to propound the following question to me, to wit, "when I throw a bowl,[17] or strike a ball with my hand, whether the motion by which the bowl or ball is moved be the hand's or the ball's own [445] motion? Or whether it be transferred out of my hand into the ball?" To which I return this short answer: that the motion by which (for example) the bowl is moved is the bowl's own motion, and not the hand's that threw it; for the hand cannot transfer its own motion, which has a material being, out of itself into the bowl, or any other thing it handles, touches, or moves; or else if it did, the hand would in a short time become weak and useless, by losing so much substance, unless new motions were as fast created, as expended.

You'll say, perhaps, that the hand and the bowl may exchange motions, as that the bowl's own motion does enter into the hand and supply that motion which went out of the hand into the bowl, by a close joining or touch, for in all things moving and moved must be a joining of the mover to the moved, either immediate or by the means of another body. I answer that this is more probable than that the hand should give out or impart motion to the bowl, and receive none from the bowl; but by reason motion cannot be transferred without matter, as being both inseparably united and but one thing, I cannot think it probable that any of the animate or self-moving matter in the hand quits the hand and

---

17. Bowl: a sphere or ball used in games of bowling or skittles.

enters into the bowl; nor that the animate matter which is in the bowl leaves the bowl and enters into the hand, because that self-moving substance is not readily prepared for so sudden a translation or transmigration.

You may say, it may as easily be done as food is received into an animal body and excrement discharged, or as air is taken in and breath sent out by the way of respiration; and that all creatures are not only produced from each other, but [446] do subsist by each other, and act by each other's assistance. I answer, it is very true that all creatures have more power and strength by a joined assistance than if every part were single and subsisted of itself. But as some parts do assist each other, so on the other side, some parts do resist each other; for though there be a unity in the nature of infinite matter, yet there are divisions also in the infinite parts of infinite matter, which causes antipathy as much as sympathy; but they being equal in assistance as well as in resistance, it causes a conformity in the whole nature of infinite matter; for if there were not contrary, or rather, I may say, different effects proceeding from the only cause, which is the only matter, there could not possibly be any, or at least so much, variety in nature as human sense and reason perceives there is.

But to return to our first argument: you may say that motion may be transferred out of one body into another without transferring any of the matter. I answer, that is impossible, unless motion were that which some call nothing, but how nothing can be transferred, I cannot imagine. Indeed, no sense and reason in nature can conceive that which is nothing; for how should it conceive that which is not in nature to be found. You'll say, perhaps, it is a substanceless thing, or an incorporeal, immaterial being or form. I answer, in my opinion, it is a mere contradiction to say a substanceless thing, form, or being, for surely in nature it cannot be. But if it be not possible that motion can be divided from matter, you may say that body from whence the motion is transferred would become less in bulk and weight, and weaker with every act of motion; [447] and those bodies into which corporeal motion or self-moving matter was received would grow bigger, heavier, and stronger. To which I answer that this is the reason which denies that there can be a translation of motion out of the moving body into the moved; for questionless, the one would grow less and the other bigger, that by losing so much substance, this by receiving.

Nay, if it were possible, as it is not, that motion could be transferred without matter, the body out of which it goes would nevertheless grow weaker; for the strength lies in the motion, unless you believe this motion which is transferred to have been useless in the mover, and only useful to the moved; or else it would be superfluous in the moved, except you say it became to be annihilated after it was transferred and had done its effect; but if so, then there would be a perpetual and infinite creation and annihilation of substanceless motion, and how there could be a creation and annihilation of nothing, my reason cannot conceive, neither is it possible, unless nature had more power than God, to create nothing and to annihilate nothing.

The truth is, it is more probable for sense and reason to believe a creation of something out of nothing than a creation of nothing out of nothing. Wherefore it cannot in sense and reason be that the motion of the hand is transferred into the bowl. But yet I do not say that the motion of the hand does not contribute to the motion of the bowl; for though the bowl has its own natural motion in itself (for nature and her creatures know of no rest, but are in a perpetual motion, though not always exterior and local, yet they have their proper and certain motions, which are not so easily [448] perceived by our grosser senses), nevertheless the motion of the bowl would not move by such an exterior local motion, did not the motion of the hand or any other exterior moving body give it occasion to move that way. Wherefore the motion of the hand may very well be said to be the cause of that exterior local motion of the bowl, but not to be the same motion by which the bowl moves. Neither is it requisite that the hand should quit its own motion because it uses it in stirring up or putting on the motion of the bowl; for it is one thing to use and another to quit; as for example, it is one thing to offer his life for his friend's service, another to employ it, and another to quit or lose it.

But, Madam, there may be infinite questions or exceptions, and infinite answers made upon one truth; but the wisest and most probable way is to rely upon sense and reason, and not to trouble the mind, thoughts, and actions of life with improbabilities, or rather impossibilities, which sense and reason knows not of nor cannot conceive. You may say a man has sometimes improbable or impossible fancies, imaginations, or chimeras in his mind, which are nothings. I answer, that those fancies and imaginations are not nothings, but as perfectly embodied as any other creatures; but by reason they are not so grossly embodied as those creatures that are composed of more sensitive and inanimate matter, man thinks or believes them to be no bodies; but were they substanceless figures, he could not have them in his mind or thoughts. The truth is, the purity of reason is not so perspicuous and plain to sense as sense is to reason, the sensitive matter being a grosser substance than the rational. And thus, Madam, [449] I have answered your proposed question, according to the ability of my reason, which I leave to your better examination, and rest in the meanwhile, Madam,

Your faithful friend and servant.

[451]   **8.**

Madam,

The other day I met with the work of that learned author Dr. Ch.,[18] which treats of natural philosophy; and amongst the rest, in the chapter of place, I found that he blames Aristotle for saying there are none but corporeal dimensions, length, breadth, and depth in nature, making, besides these corporeal, other incorporeal dimensions which he attributes to vacuum.

Truly, Madam, an incorporeal dimension or extension seems, in my opinion, a mere [452] contradiction; for I cannot conceive how nothing can have a dimension or extension, having nothing to be extended or measured. His words are these: "Imagine we therefore, that God should please to annihilate the whole stock or mass of elements, and all concretions resulting therefrom, that is, all corporeal substances now contained within the ambit or concave of the lowest heaven or lunar sphere; and having thus imagined, can we conceive that all the vast space or region circumscribed by the concave superfice[19] of the lunar sphere would not remain the same in all its dimensions, after as before the reduction of all bodies included therein to nothing?"[20]

To which, I answer, he makes nature supernatural; for although God's power may make vacuum, yet nature cannot; for God's and nature's power are not to be compared, neither is God's invisible power perceptible by nature's parts; but according to natural perception, it is impossible to conceive a vacuum, for we cannot imagine a vacuum, but we must think of a body, as your author of the circle of the moon; neither could he think of space but from one side of the circle to the other, so that in his mind he brings two sides together, and yet will have them distant; but the motions of his thoughts, being subtler and swifter than his senses, skip from side to side without touching the middle parts, like as a squirrel from bough to bough, or an ape from one table to another, without touching the ground, only cutting the air.

Next, he says that an absolute vacuum is neither an accident, nor a body, nor yet nothing, but something, because it has a being;[21] which opinion seems

---

18. Walter Charleton (1619–1707). Cavendish is discussing his *Physiologia Epicuro-Gassendo-Charltoniana: Or, A Fabrick of Science Natural, Upon the Hypothesis of Atoms* (1654). On Charleton and atomism, see the Introduction (pp. xliii–xliv).

19. Superfice: surface.

20. Charleton, *Physiologia*, book 1, chapter 6, section 1, article 4.

21. Charleton, *Physiologia*, book 1, chapter 6, section 1, article 11.

to me like that of the divine soul; but I suppose vacuum is not the [453] divine soul, nor the divine soul vacuum; or else it could not be sensible of the blessed happiness in heaven or the torments in hell.

Again he says, "Let us screw our supposition one pin higher, and farther imagine that God, after the annihilation of this vast machine, the universe, should create another in all respects equal to this, and in the same part of space wherein this now consists. First, we must conceive that, as the spaces were immense before God created the world, so also must they eternally persist of infinite extent, if he shall please at any time to destroy it. Next, that these immense spaces are absolutely immoveable."[22]

By this opinion, it seems that God's power cannot so easily make or annihilate vacuum as a substance; because he believes it to be before all matter, and to remain after all matter, which is to be eternal; but I cannot conceive why matter, or fullness of body, should not as well be infinite and eternal as his conceived vacuum; for if vacuum can have an eternal and infinite being, why may not fullness of body, or matter? But he calls vacuum immovable, which in my opinion is to make it a god; for God is only immoveable and unalterable, and this is more glorious than to be dependent upon God; wherefore, to believe matter to be eternal but yet dependent upon God is a more humble opinion than his opinion of vacuum; for if vacuum be not created, and shall not be annihilated, but is uncreated, immaterial, immoveable, infinite, and eternal, it is a god; but if it be created, God being not a creator of nothing, nor an annihilator of nothing, but of something, he cannot be a creator of vacuum; for vacuum is [454] a pure nothing. But leaving nothing to those that can make something of it, I will add no more, but rest, Madam,

Your faithful friend and servant.

9.

Madam,

That learned author of whom I made mention in my last is pleased to say in his chapter of time that time is the "twin-brother to space";[23] but if space be

---

22. Charleton, *Physiologia*, book 1, chapter 6, section 1, article 13.
23. Charleton, *Physiologia*, book 1, chapter 7, section 1, article 2.

as much as vacuum, then I say they are twin nothings; for there can be no such thing as an empty or immaterial space, but that which man calls "space" is only a distance between several corporeal parts, and time is only the variation of corporeal motions; for were there no body, there could not be any space, and were there no corporeal motion, there could not be any time.

As for time, considered in general it is nothing else but the corporeal motions in nature, and particular times are the particular corporeal motions; but duration is only a continuance or continued subsistence of the same parts, caused by the consistent motions of those parts. Neither are time, duration, place, space, magnitude, etc. dependents upon corporeal motions, but they are all one [455] and the same thing. Neither was time before, nor can be after corporeal motion, for none can be without the other, being all one. And as for eternity, it is one fixed instant, without a flux or motion.

Concerning his argument of divisibility of parts, my opinion is that there is no part in nature individable, no not that so small a part which the Epicureans name an "atom"; neither is matter separable from matter, nor parts from parts in general, but only in particulars; for though parts can be separated from parts by self-motion, yet upon necessity they must join to parts, so as there can never be a single part by itself. But hereof, as also of place, space, time, motion, figure, magnitude, etc., I have sufficiently discoursed in my former letters, as also in my book of philosophy;[24] and as for my opinion of atoms, their figures and motions (if any such things there be) I will refer you to my book of poems,[25] out of which give me leave to repeat these following lines, containing the ground of my opinion of atoms:

> All creatures, howsoever they may be named,
> Are of long, square, flat, or sharp atoms framed.[26]
> Thus several figures several tempers make.
> But what is mixed, does of the four partake.[27]
> The only cause, why things do live and die,

---

24. *Philosophical and Physical Opinions* (1663). Motion and figure are discussed throughout parts 1 and 4. For her discussion of time, see part 3, chapter 13. The preface briefly covers all of these topics.

25. I.e., *Poems, and Fancies*, originally published in 1653, with a second edition under the title *Poems and Phancies* (London: William Wilson, 1664). The passages Cavendish cites here are from the 1664 edition.

26. From "Of the Four Principal Sorts of Atoms," *Poems and Phancies* (1664), 7.

27. From "The Four Principal Figured Atoms Make the Four Elements," *Poems and Phancies* (1664), 9.

Is according as the mixed atoms lie.[28]
Thus life, and death, and young, and old,
Are as the several atoms hold;
Wit, understanding in the brain
Are as the several atoms reign;
[456] And dispositions, good, or ill,
Are as the several atoms still;
And every passion, which does rise,
Is as each several atom lies.
Thus sickness, health, and peace, and war,
Are as the several atoms are.[29]

If you desire to know more, you may read my mentioned book of poems whose first edition was printed in the year, 1653. And so taking my leave of you, I rest, Madam,

Your faithful friend and servant.

10.

Madam,

I received the book of your new author that treats of natural philosophy, which I perceive is but lately come forth;[30] but although it be new, yet there are no new opinions in it; for the author does follow the opinions of some old philosophers, and argues after the accustomed scholastical way, with hard, intricate, and nonsensical words. Wherefore I shall not take so much pains as to read it quite over, but only pick out here and there some few discourses which I shall think most convenient for the clearing of my own opinion; in the [457] number of which, is, first, that of matter, whereof the author is pleased to proclaim the opinion that holds matter to be infinite not only absurd but also impious.

---

28. A slightly altered version of lines from "What Atoms Make Life," *Poems and Phancies* (1664), 22.

29. A slightly altered version of "All Things are Governed by Atoms," *Poems and Phancies* (1664), 24.

30. Gideon Harvey, *Archelogia Philosophica Nova, or, New Principles of Philosophy* (1663). On Harvey (1636/7–1702), see the Introduction (p. xlv).

Truly, Madam, it is easily said, but hardly proved; and not to trouble you with unnecessary repetitions, I hope you do remember as yet what I have written to you in the beginning concerning the infiniteness of matter, or natural matter, where I have proved that it implies no impiety, absurdity, or contradiction at all to believe that matter is infinite; for your author's argument, concluding from the finiteness of particular creatures to nature herself,[31] is of no force; for though no part of nature is infinite in bulk, figure, or quantity, nevertheless, all the parts of infinite nature are infinite in number, which infinite number of parts must needs make up one infinite body in bulk or quantity; for as a finite body or substance is dividable into finite parts, so an infinite body, as nature or natural matter, must of necessity be dividable into infinite parts in number, and yet each part must also be finite in its exterior figure, as I have proved in the beginning by the example of a heap of grains of corn.[32]

Certainly, Madam, I see no reason, but since, according to your author, God, as the prime cause, agent, and producer of all things, and the action by which he produced all things, is infinite, the matter out of which he produced all particular creatures may be infinite also. Neither does it, to my sense and reason, imply any contradiction or impiety; for it derogates nothing from the glory and omnipotency of God, but God is still the God of nature, and nature is his servant, although infinite, depending wholly upon the will and pleasure [458] of the all-powerful God. Neither do these two infinites obstruct each other; for nature is corporeal, and God is a supernatural and spiritual infinite being, and although nature has an infinite power, yet she has but an infinite natural power, whereas God's omnipotency is infinitely exceeded beyond nature.

But your author is pleased to refute that argument which concludes from the effect to the cause, and proves matter to be infinite because God as the cause is infinite, saying that this rule does only hold in univocal things[33] (by which, I suppose, he understands things of the same kind and nature), and not in opposites. Truly, Madam, by this he limits God's power, as if God were not able to work beyond nature and natural reason or understanding; and measures God's actions according to the rules of logic, which, whether it be not more impious, you may judge yourself. And as for opposites, God and nature are not opposites, except you will call opposites those which bear a certain relation to one another, as a cause and its effect; a parent and a child; a master and a servant; and the like.

---

31. Harvey, *Archelogia*, part 2, book 1, chapter 9, section 8.
32. See in this volume, section 1, letter 2.
33. Harvey, *Archelogia*, part 2, book 1, chapter 9, section 8.

Nay, I wonder how your author can limit God's action, when as he confesses himself that the creation of the world is an infinite action. God acted finitely, says he, by an infinite action; which, in my opinion, is mere nonsense, and as much as to say a man can act weakly by a strong action, basely by an honest action, cowardly by a stout[34] action. The truth is, God, being infinite, cannot work finitely; for, as his essence, so his actions cannot have any limitation, and therefore it is most probable that God made nature infinite; for though each part of nature is finite [459] in its own figure, yet considered in general, they are infinite, as well in number as duration, except God be pleased to destroy them; nay, every particular may in a certain sense be said infinite, to wit, infinite in time or duration; for if nature be infinite and eternal, and there be no annihilation or perishing in nature, but a perpetual successive change and alteration of natural figures, then no part of nature can perish or be annihilated, and if no part of nature perishes, then it lasts infinitely in nature; that is, in the substance of natural matter; for though the corporeal motions which make the figures do change, yet the ground of the figure, which is natural matter, never changes. The same may be said of corporeal motions; for though motions change and vary infinite ways, yet none is lost in nature, but some motions are repeated again; as for example, the natural motions in an animal creature, although they are altered in the dissolution of the animal figure, yet they may be repeated again by piecemeals in other creatures; like as a commonwealth or united body in society, if it should be dissolved and dispersed, the particulars which did constitute this commonwealth or society may join to the making of another society; and thus the natural motions of a body do not perish when the figure of the body dissolves, but join with other motions to the forming and producing of some other figures.

But to return to your author, I perceive his discourse is grounded upon a false supposition, which appears by his way of arguing from the course of the stars and planets to prove the finiteness of nature; for by reason the stars and planets roll about and turn to the same point again, each [460] within a certain compass of time, he concludes nature or natural matter to be finite too. And so he takes a part for the whole, to wit, this visible world for all nature, when as this world is only a part of nature or natural matter, and there may be more and infinite worlds besides; wherefore his conclusion must needs be false, since it is built upon a false ground.

---

34. Stout: brave.

Moreover, he is as much against the eternity of matter as he is against infiniteness; concluding likewise from the parts to the whole. For, says he, since the parts of nature are subject to a beginning and ending, the whole must be too.[35] But he is much mistaken when he attributes a beginning and ending to parts, for there is no such thing as a beginning and ending in nature, neither in the whole, nor in the parts, by reason there is no new creation or production of creatures out of new matter, nor any total destruction or annihilation of any part in nature, but only a change, alteration, and transmigration of one figure into another; which change and alteration proves rather the contrary, to wit, that matter is eternal and incorruptible; for if particular figures change, they must of necessity change in the infinite matter, which itself and in its nature is not subject to any change or alteration. Besides, though particulars have a finite and limited figure and do change, yet their species do not; for mankind never changes nor ceases to be, though Peter and Paul die, or rather their figures dissolve and divide; for to die is nothing else but that the parts of that figure divide and unite into some other figures by the change of motion in those parts.

Concerning the inanimate matter, [461] which of itself is a dead, dull, and idle matter, your author denies it to be a co-agent or assistant to the animate matter; for, says he, how can dead and idle things act?[36] To which I answer that your author being, or pretending to be, a philosopher, should consider that there is difference between a principal and instrumental cause or agent; and although this inanimate or dull matter does not act of itself as a principal agent, yet it can and does act as an instrument, according as it is employed by the animate matter; for by reason there is so close a conjunction and commixture of animate and inanimate matter in nature, as they do make but one body, it is impossible that the animate part of matter should move without the inanimate; not that the inanimate has motion in herself, but the animate bears up the inanimate in the action of her own substance, and makes the inanimate work, act, and move with her, by reason of the aforesaid union and commixture.

Lastly, your author speaks much of "minimas," *viz.*, that all things may be resolved into their minimas, and what is beyond them is nothing, and that there is one maximum or biggest, which is the world, and what is beyond that, is

---

35. Harvey, *Archelogia*, part 2, book 1, chapter 9, section 8.
36. Harvey, *Archelogia*, part 2, book 1, chapter 5, section 3.

infinite.[37] Truly, Madam, I must ingeniously[38] confess, I am not so high learned as to penetrate into the true sense of these words; for he says they are both divisible and indivisible, and yet no atoms, which surpasses my understanding; for there is no such thing as biggest and smallest in nature, or in the infinite matter; for who can know how far this world goes or what is beyond it? There may be infinite worlds, as I said before, for aught we know; for God and nature cannot be [462] comprehended, nor their works measured; if we cannot find out the nature of particular things which are subject to our exterior senses, how shall we be able to judge of things not subject to our senses?

But your author does speak so presumptuously of God's actions, designs, decrees, laws, attributes, powers, and secret counsels, and describes the manner how God created all things, and the mixture of the elements to a hair, as if he had been God's counselor and assistant in the work of creation; which whether it be not more impiety than to say matter is infinite, I'll let others judge. Neither do I think this expression to be against the holy Scripture; for though I speak as a natural philosopher, and am unwilling to cite the Scripture, which only treats of things belonging to faith, and not to reason, yet I think there is not any passage which plainly denies matter to be infinite and eternal, unless it be drawn by force to that sense. Solomon says "that there is not anything new";[39] and in another place it is said "that God is all fulfilling";[40] that is, the will of God is the fulfilling of the actions of nature. Also the Scripture says "that God's ways are unsearchable, and past finding out."[41] Wherefore, it is easier to treat of nature than the God of nature; neither should God be treated of by vain philosophers, but by holy divines, which are to deliver and interpret the word of God without sophistry, and to inform us as much of God's works as he has been pleased to declare and make known. And this is the safest way, in the opinion of, Madam,

Your faithful friend and servant.

---

37. Harvey, *Archelogia*, part 2, book 1, chapter 6, section 5.
38. Ingeniously: ingenuously, straightforwardly.
39. Eccles. 1:9.
40. Perhaps a reference to Eph. 1:23, "him that filleth in all things."
41. Rom. 11:33.

22.

Madam,

You were pleased to desire my opinion of the works of that learned and ingenious writer B.[42] Truly, Madam, I have read but some part of his works; but as much as I have read, I have observed, he is a very civil, eloquent, and rational writer; the truth is, his style is a gentleman's style. And in particular, concerning his experiments, I must needs say this, that, in my judgment, he has expressed himself [496] to be a very industrious and ingenious person; for he does neither puzzle nature, nor darken truth with hard words and compounded languages, or nice distinctions; besides, his experiments are proved by his own action.

But give me leave to tell you, that I observe he studies the different parts and alterations more than the motions which cause the alterations in those parts; whereas, did he study and observe the several and different motions in those parts, how they change in one and the same part, and how the different alterations in bodies are caused by the different motions of their parts, he might arrive to a vast knowledge by the means of his experiments; for certainly experiments are very beneficial to man.

In the next place, you desire my opinion of the book called *The Discourses of the Virtuosi in France*.[43] I am sorry, Madam, this book comes so late to my hands that I cannot read it so slowly and observingly as to give you a clear judgment of their opinions or discourses in particular; however, in general, and for what I have read in it, I may say, it expresses the French to be very learned and eloquent writers, wherein I thought our English had exceeded them, and that they did only excel in wit and ingenuity; but I perceive most nations have of all sorts. The truth is, ingenious and subtle wit brings news; but learning and experience brings proofs, at least, argumental discourses; and the French are much to be commended that they endeavor to spend their time wisely, honorably, honestly, and profitably, not only for the good and benefit of their own, but also of other nations.

But before I conclude, give me leave to tell you that, concerning the curious and profitable [497] arts mentioned in their discourses, I confess I do much admire them, and partly believe they may arrive to the use of many of them; but

---

42. A reference to Robert Boyle (1627–1691). See the Introduction (p. xlvi).

43. Cavendish is referring to *A General Collection of Discourses of the Virtuosi of France, upon Questions of All Sorts of Philosophy and other Natural Knowledge* [...] *Rendered into English by G. Havers* (1664). On this book, see the Introduction (p. xlvi).

there are two arts which I wish with all my heart I could obtain. The first is to argue without error in all kinds, modes, and figures in a quarter of an hour; and the other is to learn a way to understand all languages in six hours. But as for the first, I fear, if I want a thorough understanding in every particular argument, cause, or point, a general art or mode of words will not help me, especially if I, being a woman, should want discretion. And as for the second, my memory is so bad, that it is beyond the help of art, so that nature has made my understanding harder or closer than glass, through which the sun of verity cannot pass, although its light does; and therefore I am confident I shall not be made or taught to learn this mentioned art in six hours, no not in six months. But I wish all arts were as easily practiced as mentioned; and thus I rest, Madam,

Your faithful friend and servant.

[500]   **24.**

Madam,

I have heard that artists do glory much in their glasses,[44] tubes, engines, and stills, and hope by their glasses and tubes to see invisible things, and by their engines to produce incredible effects, and by their stills, fire, and furnaces, to create as nature does; but all this is impossible to be done. For art cannot arrive to that degree as to know perfectly nature's secret and fundamental actions, her purest matter, and subtlest motions; and it is enough if artists can but produce such things as are for man's conveniencies and use, although they never can see the smallest or rarest bodies, nor great and vast bodies at a great distance, nor make or create a vegetable, animal, or the like, as nature does; for nature, being infinite, has also infinite [501] degrees of figures, sizes, motions, densities, rarities, knowledge, etc., as you may see in my book of philosophy,[45] as also in my book of poems, especially that part that treats of little, minute creatures, which I there do name, for want of other expressions, "fairies";[46] for I have considered much the several sizes of creatures, although I gave it out but for a fancy in the

---

44. Glasses: optical instruments, including telescopes and microscopes.
45. See *Philosophical and Physical Opinions* (1663), part 1, chapters 7–9.
46. Both the 1653 and 1664 editions of *Poems, and Fancies* include poems about fairies, such as "The Fairy Queen" and "The Pastime of the Queen of Fairies."

mentioned book, lest I should be thought extravagant to declare that conception of mine for a rational truth.

But if some small bodies cannot be perfectly seen but by the help of magnifying glasses and such as they call "microscopia," I pray, nature being infinite, what figures and sizes may there not be, which our eyes with all the help of art are not capable to see? For certainly, nature has more curiosities than our exterior senses, helped by art, can perceive. Wherefore I cannot wonder enough at those that pretend to know the least or greatest parts or creatures in nature, since no particular creature is able to do it.

But concerning artists, you would fain know, Madam, whether the artist is beholden to the conceptions of the student? To which I return this short answer: that, in my judgment, without the student's conceptions, the artist could not tell how to make experiments. The truth is, the conceptions of studious men set the artists on work, although many artists do ungratefully attribute all to their own industry. Neither does it always belong to the studious conceptor to make trials or experiments, but he leaves that work to others, whose time is not so much employed with thoughts or speculations as with actions, for [502] the contemplator is the designer, and the artist the workman or laborer, who ought to acknowledge him his master, as I do your Ladyship, for I am in all respects, Madam,

Your Ladyship's humble and faithful servant.

25.

Madam,

Your command in your last was to send you my opinion concerning the division of religions, or of the several opinions in religions; I suppose you mean the division of the religion, not of religions; for certainly there is but one divine truth, and consequently but one true religion. But natural men being composed of many diverse parts, as of several motions and figures, have diverse and several ideas, which the grosser corporeal motions conceive to be diverse and several gods, as being not capable to know the great and incomprehensible God, who is above nature. For example: do but consider, Madam, what strange opinions the heathens had of God, and how they divided him into so many several persons, with so many several bodies, like men; whereas, surely God considered in his essence, he being a spirit, as the Scripture describes him, can neither have soul nor body, as he [503] is a God, but is an immaterial being; only the heathens did conceive him to have parts, and so divided the incomprehensible God into several deities, at least they had several deitical ideas, or rather fancies of him. But, Madam, I confess my ignorance in this great mystery, and honor and praise the omnipotent, great, and incomprehensible God with all fear and humility as I ought; beseeching his infinite mercy to keep me from such presumption whereby I might profane his holy name, and to make me obedient to the Church, as also to grant me life and health, that I may be able to express how much I am, Madam,

Your faithful friend and servant.

26.

Madam,

Since I spoke of religion in my last, I cannot but acquaint you that I was the other day in the company of Sir. P. H. and Sir R. L.,[47] where amongst other discourses they talked of predestination and freedom. Sir P. H. accounted the opinion of predestination not only absurd but blasphemous; for, said he, predestination makes God appear cruel, as first to create angels and man, and then to make them fall from their glory, and damn them eternally. For God, said he, [504] knew before he made them, they would fall. Neither could he imagine from whence that pride and presumption did proceed, which was the cause of the angels' fall, for it could not proceed from God, God being infinitely good.

Sir R. L. answered that this pride and presumption did not come from God, but from their own nature. But, replied Sir P. H., God gave them that nature, for they had it not of themselves, but all what they were, their essence and nature, came from God the creator of all things, and to suffer that which was in his power to hinder was as much as to act.

Sir R. L. said, God gave both angels and man a free will at their creation. Sir P. H. answered that a free will was a part of a divine attribute, which surely God would not give away to any creature. Next, said he, he could not conceive why God should make creatures to cross and oppose him; for it were neither an act of wisdom to make rebels, nor an act of justice to make devils; so that neither in

---

47. On identifying these men, see the Introduction (p. xlvii).

his wisdom, justice, nor mercy, God could give leave that angels and man should fall through sin; neither was God ignorant that angels and man would fall; for surely, said he, God knew all things, past, present, and to come; wherefore, said he, free will does weaken the power of God, and predestination does weaken the power of man, and both do hinder each other. Besides, said he, since God did confirm the rest of the angels in the same state they were before, so as they could not fall afterwards, he might as well have created them all so at first.

But Sir R. L. replied that God suffered angels and man to fall for his glory, to show his [505] justice in devils and his mercy in man; and that the devils expressed God's omnipotency as much as the blessed. To which Sir P. H. answered that they expressed more God's severity in those horrid torments they suffer through their natural imperfections than his power in making and suffering them to sin.

Thus they discoursed. And to tell you truly, Madam, my mind was more troubled than delighted with their discourse; for it seemed rather to detract from the honor of the great God than to increase his glory; and no creature ought either to think or to speak anything that is detracting from the glory of the creator. Wherefore I am neither for predestination, nor for an absolute free will, neither in angels, devils, nor man; for an absolute free will is not competent to any creature; and although nature be infinite, and the eternal servant to the eternal and infinite God, and can produce infinite creatures, yet her power and will is not absolute, but limited; that is, she has a natural free will, but not a supernatural, for she cannot work beyond the power God has given her. But those mystical discourses belong to divines, and not to any layperson, and I confess myself very ignorant in them. Wherefore I will nor dare not dispute God's actions, being all infinitely wise, but leave that to divines, who are to inform us what we ought to believe and how we ought to live. And thus taking my leave of you for the present, I rest, Madam,

Your faithful friend and servant.

[506]　　　　　　　　　　　　　**27.**

Madam,

You are pleased to honor me so far that you do not only spend some time in the perusing of my book called *Philosophical Opinions*, but take it so much into your consideration as to examine every opinion of mine which dissents from the common way of the schools, marking those places which seem somewhat obscure, and desiring my explanation of them; all which, I do not only acknowledge as a great favor, but as an infallible testimony of your true and unfeigned friendship; and I cannot choose but publish it to all the world; both for the honor of yourself, as to let everybody know the part of so true a friend, who is so much concerned for the honor and benefit of my poor works; as also for the good of my mentioned book, which by this means will be rendered more intelligible; for I must confess that my *Philosophical Opinions* are not so plain and perspicuous as to be perfectly understood at the first reading, which I am sorry for. And there be two chief reasons why they are so: first, because they are new, and never vented before; for they have their original merely from my own conceptions, and are not taken out of other philosophers. Next, because I, being a woman, and not bred up to scholarship, did want names and terms of art, and therefore being not versed in the writings of other philosophers [507] but what I knew by hearing, I could not form my named book so methodically, and express my opinions so artificially and clearly as I might have done, had I been studious in the reading of philosophical books, or bred a scholar; for then I might have dressed them with a fine colored covering of logic and geometry, and set them out in a handsome array; by which I might have also covered my ignorance, like as stage-players do cover their mean persons or degrees with fine clothes.

But, as I said, I, being void of learning and art, did put them forth according to my own conceptions, and as I did understand them myself; but since I have hitherto, by the reading of those famous and learned authors you sent me, attained to the knowledge of some artificial terms, I shall not spare any labor and pains to make my opinions so intelligible, that everyone, who without partiality, spleen,[48] or malice, does read them, may also easily understand them. And thus I shall likewise endeavor to give such answers to your scruples, objections, or questions as may explain those passages which seem obscure, and satisfy your desire.

In the first place, and in general, you desire to know, "whether any truth may be had in natural philosophy." For since all this study is grounded upon probability, and he that thinks he has the most probable reasons for his opinion may be as far off from truth as he who is thought to have the least; nay, what seems most probable today may seem least probable tomorrow, especially if an ingenious opposer bring rational arguments against it. Therefore you think it is

---

48. Spleen: spite.

but vain for anyone to trouble his brain with searching and enquiring after such things wherein neither truth nor certainty can [508] be had.

To which I answer that the undoubted truth in natural philosophy is, in my opinion, like the philosopher's stone in chemistry, which has been sought for by many learned and ingenious persons, and will be sought as long as the art of chemistry does last; but although they cannot find the philosopher's stone, yet by the help of this art they have found out many rare things both for use and knowledge. The like in natural philosophy, although natural philosophers cannot find out the absolute truth of nature, or nature's ground-works, or the hidden causes of natural effects; nevertheless they have found out many necessary and profitable arts and sciences to benefit the life of man; for without natural philosophy we should have lived in dark ignorance, not knowing the motions of the heavens, the cause of the eclipses, the influences of the stars, the use of numbers, measures, and weights, the virtues and effects of vegetables and minerals, the art of architecture, navigation, and the like. Indeed all arts and sciences do ascribe their original to the study of natural philosophy; and those men are both unwise and ungrateful that will refuse rich gifts because they cannot be masters of all wealth; and they are fools that will not take remedies when they are sick because medicines can only recover them from death for a time, but not make them live forever.

But to conclude, probability is next to truth, and the search for a hidden cause finds out visible effects; and this truth do natural philosophers find, that there are more fools than wise [509] men, which fools will never attain to the honor of being natural philosophers. And thus leaving them, I rest, Madam,

Your Ladyship's humble and faithful servant.

28.

Madam,

Your desire is to know, since I say nature is wise, whether all her parts must be wise also? To which, I answer that (by your favor) all her parts are not fools; but yet it is no necessary consequence that because nature is infinitely wise, all her parts must be so too, no more than if I should say, nature is infinite, therefore every part must be infinite. But it is rather necessary that because nature is infinite, therefore not any single part of hers can be infinite, but must be finite.

Next, you desire to know, whether nature or the self-moving matter is subject to err and to commit mistakes? I answer, although nature has naturally an infinite wisdom and knowledge, yet she has not a most pure and entire perfection, no more than she has an absolute power; for a most pure and entire perfection belongs only to God; and though she is infinitely naturally wise in herself, yet her parts or particular creatures may commit errors and mistakes; [510] the truth is, it is impossible but that parts or particular creatures must be subject to errors, because no part can have a perfect or general knowledge, as being but a part and not a whole; for knowledge is in parts, as parts are in matter. Besides, several corporeal motions, that is, several self-moving parts, do delude and oppose each other by their opposite motions; and this opposition is very requisite in nature to keep a mean and hinder extremes; for were there not opposition of parts, nature would run into extremes, which would confound her and all her parts.

And as for delusion, it is part of nature's delight, causing the more variety; but there be some actions in nature which are neither perfect mistakes, nor delusions, but only want of a clear and thorough perception; as for example, when a man is sailing in a ship, he thinks the shore moves from the ship, when as it is the ship that moves from the shore; also when a man is going backward from a looking-glass, he thinks the figure in the glass goes inward, whereas it is himself that goes backward, and not his figure in the glass. The cause of it is that the perception in the eye perceives the distanced body, but not the motion of the distance or medium; for though the man may partly see the motion of the visible parts, yet he does not see the parts or motion of the distance or medium, which is invisible, and not subject to the perception of sight. And since a pattern cannot be made if the object be not visible, hence I conclude that the motion of the medium cannot make perception, but that it is the perceptive motions of the eye which pattern out an object as it is visibly presented to the corporeal motions in the eye; for according as the object is [511] presented, the pattern is made, if the motions be regular. For example, a fired end of a stick, if you move it in a circular figure, the sensitive corporeal motions in the eye pattern out the figure of fire, together with the exterior or circular motion, and apprehend it as a fiery circle; and if the stick be moved any otherwise, they pattern out such a figure as the fired end of the stick is moved in; so that the sensitive pattern is made according to the exterior corporeal figurative motion of the object, and not according to its interior figure or motions. And this, Madam, is in short my answer to your propounded questions, by which, I hope, you understand plainly the meaning of, Madam,

Your faithful friend and servant.

29.

Madam,

The scruples or questions you sent me last are these following. First, you desire to be informed what I mean by "phantasms" and "ideas." I answer, they are figures made by the purest and subtlest degree of self-moving matter, that is to say, by the rational corporeal motions, and are the same with thoughts or conceptions.

Next, your question is, what do I understand by "sensitive life"? I answer, [512] it is that part of self-moving matter which in its own nature is not so pure and subtle as the rational, for it is but the laboring, and the rational is the designing part of matter.

Your third question is, "whether this sensitive self-moving matter be dense or rare?" I answer, density and rarity are only effects caused by the several actions, that is, the corporeal motions of nature; wherefore it cannot properly be said that sensitive matter is either dense or rare; for it has a self-power to contract and dilate, compose and divide, and move in any kind of motion whatsoever, as is requisite to the framing of any figure; and thus I desire you to observe well, that when I say the rational part of matter is purer in its degree than the sensitive, and that this is a rare and acute matter, I do not mean that it is thin like a rare egg, but that it is subtle and active, penetrating and dividing, as well as dividable.

Your fourth question is, "what this sensitive matter works upon?" I answer, it works with and upon another degree of matter, which is not self-moving, but dull, stupid, and immoveable in its own nature, which I call the inanimate part or degree of matter.

Your fifth question is, "whether this inanimate matter do never rest?" I answer, it does not; for the self-moving matter being restless in its own nature, and so closely united and commixed with the inanimate as they do make but one body, will never suffer it to rest; so that there is no part in nature but is moving;

the animate matter in itself, or its own nature, the inanimate by the help or means of the animate.

Your sixth question is, "If there be a thorough mixture of the parts of animate and inanimate matter, whether those parts do retain each their own [513[49]] nature and substance, so that the inanimate part of matter remains dull and stupid in its essence and nature, and the animate full of self-motion, or all self-motion?" I answer, although every part and particle of each degree are closely intermixed, nevertheless this mixture does not alter the interior nature of those parts or degrees; as for example, a man is composed of soul and body, which are several parts, but joined as into one substance, *viz.*, man, and yet they retain each their own proprieties and natures; for although soul and body are so closely united as they do make but one man, yet the soul does not change into the body, nor the body into the soul, but each continues in its own nature as it is. And so likewise in infinite matter, although the degrees or parts of matter are so thoroughly intermixed as they do make but one body or substance, which is corporeal nature, yet each remains in its nature as it is, to wit, the animate part of matter does not become dull and stupid in its nature, but remains self-moving; and the inanimate, although it does move by the means of the animate, yet it does not become self-moving, but each keeps its own interior nature and essence in their commixture. The truth is, there must of necessity be degrees of matter, or else there would be no such various and several effects in nature, as human sense and reason do perceive there are; and those degrees must also retain each their own nature and proprieties, to produce those various and curious effects. Neither must those different degrees vary or alter the nature of infinite matter; for matter must and does continue one and the same in its nature, that is, matter cannot be divided from [514] being matter.

And this is my meaning when I say in *Philosophical Opinions*, "there is but one kind of matter."[50] Not that matter is not dividable into several parts or degrees, but I say, although matter has several parts and degrees, yet they do not alter the nature of matter, but matter remains one and the same in its own kind, that is, it continues still matter in its own nature notwithstanding those degrees; and thus I do exclude from matter all that which is not matter, and do firmly believe that there can be no commixture of matter and no matter in nature; for this would breed a mere confusion in nature.

---

49. Pages 513–16 are misnumbered in Cavendish's original, as 515, 516, 513, and 514, respectively.

50. See *Philosophical and Physical Opinions* (1663), part 1, chapters 1–2.

Your seventh question is, "whether that which I name the rational part of self-moving matter makes as much variety as the sensitive?" To which I answer that, to my sense and reason, the rational part of animate or self-moving matter moves not only more variously, but also more swiftly than the sensitive; for thoughts are sooner made than words spoke, and a certain proof of it are the various and several imaginations, fancies, conceptions, memories, remembrances, understanding, opinions, judgments, and the like; as also the several sorts of love, hate, fear, anger, joy, doubt; and the like passions.

Your eighth question is, "whether the sensitive matter can and does work in itself and in its own substance and degree?" My answer is that there is no inanimate matter without animate, nor no animate without inanimate, both being so curiously and subtly intermixed as they make but one body. Nevertheless the several parts of this one body may move several ways. Neither are the several degrees bound to an equal mixture, no more than the several parts [515] of one body are bound to one and the same size, bigness, shape, or motion; or the sea is bound to be always at the high tide; or the moon to be always at the full; or all the veins or brains in animal bodies are bound to be of equal quantity; or every tree of the same kind to bear fruit, or have leaves of equal number; or every apple, pear, or plum, to have an equal quantity of juice; or every bee to make as much honey and wax as the other.

Your ninth question is, "whether the sensitive matter can work without taking patterns?" My answer is that all corporeal motion is not patterning, but all patterning is made by corporeal motion; and there be more several sorts of corporeal motions than any single creature is able to conceive, much less to express. But the perceptive corporeal motions are the ground-motions in nature, which make, rule, and govern all the parts of nature, as to move to production, or generation, transformation, and the like.

Your tenth question is, "how is it possible that numerous figures can exist in one part of matter? For it is impossible that two things can be in one place, much less many." My answer in short is that it were impossible, were a part of matter and the numerous figures several and distinct things; but all is but one thing, that is, a part of matter moving variously; for there is neither magnitude, place, figure, nor motion in nature but what is matter or body. Neither is there any such thing as time. Wherefore it cannot properly be said, "there was," and "there shall be"; but only "there is." Neither can it properly be said, from this to that place; but only in [516] reference to the several moving parts of the only infinite matter. And thus much to your questions; I add no more, but rest, Madam,

Your faithful friend and humble servant.

30.

Madam,

In your last, you were pleased to express that some men, who think themselves wise, did laugh in a scornful manner at my opinion, when I say that every creature has life and knowledge, sense and reason; counting it not only ridiculous, but absurd; and asking, whether you did or could believe a piece of wood, metal, or stone had as much sense as a beast, or as much reason as a man, having neither brain, blood, heart, nor flesh; nor such organs, passages, parts, nor shapes as animals.

To which I answer that it is not any of these mentioned things that makes life and knowledge, but life and knowledge is the cause of them, which life and knowledge is animate matter, and is in all parts of all creatures. And to make it more plain and perspicuous, human sense and reason may perceive that wood, stone, or metal acts as wisely as an animal; as for example, rhubarb or the like drugs will act very wisely in purging;[51] and antinomy or the [517] like will act very wisely in vomiting; and opium will act very wisely in sleeping; also quicksilver or mercury will act very wisely, as those that have the French disease[52] can best witness. Likewise the lodestone acts very wisely, as mariners or navigators will tell you. Also wine made of fruit, and ale of malt, and distilled aqua vitae will act very subtly; ask the drunkards, and they can inform you.

Thus infinite examples may be given, and yet man says all vegetables and minerals are insensible and irrational, as also the planets and elements; when as yet the planets move very orderly and wisely, and the elements are more active, nay, more subtle and searching than any of the animal creatures; witness fire, air, and water. As for the earth, she brings forth her fruit, if the other elements do not cause abortives, in due season; and yet man believes vegetables, minerals, and elements are dead, dull, senseless, and irrational creatures, because they have not such shapes, parts, nor passages as animals, nor such exterior and local motions as animals have. But man does not consider the various, intricate, and obscure ways of nature, unknown to any particular creature; for what our senses are not capable to know, our reason is apt to deny.

Truly, in my opinion, man is more irrational than any of those creatures when he believes that all knowledge is not only confined to one sort of creatures,

---

51. Rhubarb was used for medicinal purposes due to its laxative effect.

52. The "French disease" was syphilis, which was sometimes treated with ointments made with mercury.

but to one part of one particular creature, as the head or brain of man; for who can in reason think that there is no other sensitive and rational knowledge in infinite matter, but what is only in man or animal creatures? It is a [518] very simple and weak conclusion to say other creatures have no eyes to see, no ears to hear, no tongues to taste, no noses to smell, as animals have; wherefore they have no sense or sensitive knowledge; or because they have no head nor brain as man has, therefore they have no reason, nor rational knowledge at all; for sense and reason, and consequently sensitive and rational knowledge, extends further than to be bound to the animal eye, ear, nose, tongue, head, or brain; but as these organs are only in one kind of nature's creatures, as animals, in which organs the sensitive corporeal motions make the perception of exterior objects, so there may be infinite other kinds of passages or organs in other creatures unknown to man, which creatures may have their sense and reason, that is, sensitive and rational knowledge, each according to the nature of its figure; for as it is absurd to say that all creatures in nature are animals, so it is absurd to confine sense and reason only to animals; or to say that all other creatures, if they have sense and reason, life and knowledge, it must be the same as is in animals.

I confess, it is of the same degree, that is, of the same animate part of matter, but the motions of life and knowledge work so differently and variously in every kind and sort, nay, in every particular creature, that no single creature can find them out. But, in my opinion, not any creature is without life and knowledge, which life and knowledge is made by the self-moving part of matter, that is, by the sensitive and rational corporeal motions. And as it is no consequence that all creatures must be alike in their exterior shapes, figures, and motions, because they are all produced out of one and the [519] same matter, so neither does it follow that all creatures must have the same interior motions, natures, and proprieties, and so consequently the same life and knowledge, because all life and knowledge is made by the same degree of matter, to wit, the animate.

Wherefore though every kind or sort of creatures has different perceptions, yet they are not less knowing; for vegetables, minerals, and elements may have as numerous and as various perceptions as animals, and they may be as different from animal perceptions as their kinds are; but a different perception is not therefore no perception. Neither is it the animal organs that make perception, nor the animal shape that makes life, but the motions of life make them.

But some may say, it is irreligious to believe any creature has rational knowledge but man. Surely, Madam, the God of nature, in my opinion, will be adored by all creatures, and adoration cannot be without sense and knowledge. Wherefore it is not probable that only man, and no creature else, is capable to adore and worship the infinite and omnipotent God, who is the God of nature and of all creatures. I should rather think it is irreligious to confine sense and reason only to man, and to say that no creature adores and worships God but man; which, in my judgment, argues a great pride, self-conceit, and presumption. And thus, Madam, having declared my opinion plainly concerning this subject, I will detain you no longer at this present, but rest, Madam,

Your constant friend and faithful servant.

[520] **31.**

Madam,

[...] [521] [...] Next, you desire my opinion of "vacuum," whether there be any, or not? For you say I determine nothing of it in my book of *Philosophical Opinions.*[53] Truly, Madam, my sense and reason cannot believe a vacuum, because there cannot be an empty nothing; but change of motion makes all the alteration of figures, and consequently all that which is called "place," "magnitude," "space," and the like; for matter, motion, figure, place, magnitude, etc. are but one thing. But some men, perceiving the alteration but not the subtle motions, believe that bodies move into each other's place, which is impossible, because several places are only several parts, so that, unless one part could make itself another part, no part can be said to succeed into another's place, but it is impossible that one part should make itself another part, for it cannot be another and itself, no more than nature can be nature and not nature; [522] wherefore change of place is only change of motion, and this change of motion makes alteration of figures.

Thirdly, you say you cannot understand what I mean by "creation," for you think that creation is a production or making of something out of nothing. To tell you really, Madam, this word is used by me for want of a better expression; and I do not take it in so strict a sense as to understand by it a divine or super-natural creation, which only belongs to God, but a natural creation, that is, a natural production or generation, for nature cannot create or produce something out of nothing. And this production may be taken in a double sense. First,

---

53. *Philosophical and Physical Opinions* (1663) does include a chapter entitled "Of Vacuum" (part 1, chapter 6), but it is, as Cavendish suggests here, inconclusive; she writes that "to treat of vacuum, whether there be any or not, is very difficult, for there is as much to be said of one side as of the other."

in general, as for example, when it is said that all creatures are produced out of infinite matter; and in this respect every particular creature which is finite, that is, of a circumscribed and limited figure, is produced of infinite matter, as being a part thereof. Next, "production" is taken in a more strict sense, to wit, when one single creature is produced from another; and this is either generation properly so called, as when in every kind and sort each particular produces its like, or it is such a generation whereby one creature produces another, each being of a different kind or species, as for example, when an animal produces a mineral, as when a stone is generated in the kidneys, or the like; and in this sense one finite creature generates or produces another finite creature, the producer as well as the produced being finite; but in the first sense finite creatures are produced out of infinite matter.

Fourthly, you confess you cannot well apprehend [523] my meaning when I say that the several kinds are as infinite as the particulars;[54] for your opinion is that the number of particulars must needs exceed the number of kinds. I answer: I mean in general the infinite effects of nature which are infinite in number, and the several kinds or sorts of creatures are infinite in duration, for nothing can perish in nature.

Fifthly, when I say that ascending and descending is often caused by the exterior figure or shape of a body,[55] witness a bird, who although he is of a much bigger size and bulk than a worm, yet can by his shape lift himself up more agilely and nimbly than a worm. Your opinion is that his exterior shape does not contribute anything towards his flying, by reason a bird being dead retains the same shape, but yet cannot fly at all.

But, truly, Madam, I would not have you think that I do exclude the proper and interior natural motion of the figure of a bird, and the natural and proper motions of every part and particle thereof; for that a bird when dead keeps his shape and yet cannot fly, the reason is that the natural and internal motions of the bird and the bird's wings are altered towards some other shape or figure, if not exteriously, yet interiously; but yet the interior natural motions could not effect any flying or ascending without the help of the exterior shape; for a man or any other animal may have the same interior motions as a bird has, but wanting such an exterior shape, he cannot fly; whereas had he wings like a bird, and the interior natural motions of those wings, he might without doubt fly as well as a bird does.

---

54. *Philosophical and Physical Opinions* (1663), part 4, chapter 10.
55. *Philosophical and Physical Opinions* (1663), part 4, chapter 20.

Sixthly, concerning the descent of heavy bodies, [524] that it is more forcible than the ascent of light bodies, you do question the truth of this my opinion. Certainly, Madam, I cannot conceive it to be otherwise by my sense and reason; for though fire that is rare does ascend with an extraordinary quick motion, yet this motion is, in my opinion, not so strong and piercing as when grosser parts of creatures do descend; but there is difference in strength and quickness; for had not water a stronger motion and another sort of figure than fire, it could not suppress fire, much less quench it. But smoke, which is heavier than flame, flies up, or rises before, or rather, above it. Wherefore I am still of the same opinion that heavy bodies descend more forcibly than light bodies do ascend, and it seems most rational to me.

Lastly, I perceive you cannot believe that all bodies have weight; by reason, if this were so, the sun and the stars would have long since covered the earth. In answer to this objection, I say that as there can be no body without figure and magnitude, so consequently not without weight, were it no bigger than an atom; and as for the sun and the stars not falling down, or rising higher, the reason is not their being without weight, but their natural and proper motion, which keeps them constantly in their spheres; and it might as well be said a man lives not, or is not, because he does not fly like a bird, or dive and catch fish like a cormorant, or dig and undermine like a mole, for those are motions not proper to his nature.

And these, Madam, are my answers to your objections, which if they do satisfy you, it is all I desire; if not, I shall endeavor hereafter to make my meaning more [525] intelligible, and study for other more rational arguments than these are, to let you see how much I value both the credit of my named book, and your Ladyship's commands; which assure yourself, shall never be more faithfully performed, than by, Madam,

Your Ladyship's most obliged friend and humble servant.

32.

Madam,

Since my opinion is that the animate part of matter, which is sense and reason, life and knowledge, is the designer, architect, and creator of all figures in nature; you desire to know, whence this animate matter, sense and reason, or life and knowledge (call it what you will, for it is all one and the same thing) is produced? I answer, it is eternal.

But then you say, it is coequal with God. I answer, that cannot be. For God is above all natural sense and reason, which is natural life and knowledge; and therefore it cannot be coequal with God, except it be meant in eternity, as being without beginning and end. But if God's power can make man's soul, as also the good and evil spirits to last eternally without end, he may by his omnipotency make as well things without beginning.

You will say, if nature were eternal, it [526] could not be created, for the word "creation" is contrary to eternity. I answer, Madam, I am no scholar for words; for if you will not use the word "creation," you may use what other word you will; for I do not stand upon nice words and terms, so I can but express my conceptions. Wherefore, if it be (as in reason it cannot be otherwise) that nothing in nature can be annihilated, nor anything created out of nothing, but by God's special and all-powerful decree and command, then nature must be as God has made her, until he destroy her. But if nature be not eternal, then the gods of the heathens were made in time, and were no more than any other creature which is as subject to be destroyed as created; for they conceived their gods, as we do men, to have material bodies but an immaterial spirit, or as some learned men imagine, to be an immaterial spirit, but to take several shapes, and so to perform several corporeal actions; which truly is too humble and mean a conception of an immaterial being, much more of the great and incomprehensible God; which I do firmly believe is a most pure, all-powerful immaterial being, which does all things by his own decree and omnipotency without any corporeal actions or shapes, such as some [people] fancy demons and the like spirits.

But to return to the former question; you might as well enquire how the world or any part of it was created, or how the variety of creatures came to be, as ask how reason and sensitive corporeal knowledge was produced from God; but after what manner or way, is impossible for any creature or part of nature to know, [527] for God's ways are incomprehensible and supernatural. And thus much I believe, that as God is an eternal creator, which no man can deny, so he has also an eternal creature, which is nature or natural matter.

But put the case nature or natural matter was made when the world was created, might not God give this natural matter self-motion as well as he gave self-motion to spirits and souls? And might not God endue this matter with sense and reason, as well as he endued man? Shall or can we bind up God's actions with our weak opinions and foolish arguments? Truly, if God could not act more than man is able to conceive, he were not a God of an infinite power;

but God is omnipotent, and his actions are infinite, supernatural, and past finding out; wherefore he is rather to be admired, adored, and worshipped, than to be ungloriously discoursed of by vain and ambitious men, whose foolish pride and presumption drowns their natural judgment and reason; to which leaving them, I rest, Madam,

Your faithful friend and servant.

[528]   **33.**

Madam,

In obedience to your commands, I here send you also an explanation and clearing of those places and passages in my book of philosophy which in your last letter you were pleased to mark, as containing some obscurity and difficulty of being understood.

First, when I say, "nature is an individable matter,"[56] I do not mean as if nature were not dividable into parts; for because nature is material, therefore she must also needs be dividable into parts. But my meaning is that nature cannot be divided from matter, nor matter from nature, that is, nature cannot be immaterial, nor no part of nature, but if there be anything immaterial, it does not belong to nature. Also, when I call nature a "multiplying figure,"[57] I mean that nature makes infinite changes, and so infinite figures.

Next, when I say, "there are infinite divisions in nature,"[58] my meaning is not that there are infinite divisions of one single part, but that infinite matter has infinite parts, sizes, figures, and motions, all being but one infinite matter, or corporeal nature. Also when I say "single parts," I mean not parts subsisting by themselves, précised[59] from each other, but single, that is, several or different, by reason of their different figures. Likewise, when I name "atoms," I mean small parts of matter; and when I speak of place [529] and time, I mean only the variation of corporeal figurative motions.

---

56. *Philosophical and Physical Opinions* (1663), part 3, chapter 13.
57. *Philosophical and Physical Opinions* (1663), part 3, chapter 13.
58. *Philosophical and Physical Opinions* (1663), part 1, chapter 11.
59. Précised: particularized, separated individually.

Again, when I say, "nature has not an absolute power, because she has an infinite power,"[60] I mean by "absolute" as much as finite, or circumscribed; and in this sense nature cannot have an absolute power, for the infiniteness hinders the absoluteness; but when in my former letters I have attributed an absolute power only to God, and said that nature has not an absolute power, but that her power, although it be infinite, yet cannot extend beyond nature, but is an infinite natural power; I understand by an absolute power, not a finite power, but such a power which only belongs to God, that is, a supernatural and divine power, which power nature cannot have, by reason she cannot make any part of her body immaterial, nor annihilate any part of her creatures, nor create any part that was not in her from eternity, nor make herself a deity; for though God can empower her with a supernatural gift, and annihilate her when he pleases, yet she is no ways able to do it herself. [. . .]

Also, when I say that "matter would have power over infinite, and infinite over matter, and eternal over both,"[61] I mean that some corporeal actions endeavor [530] to be more powerful than others, and thus the whole strives to overpower the parts, and the parts the whole; as for example, if one end of a string were tied about the little finger of one's hand, and the other end were in the power of the other whole hand, and both did pull several and opposite ways; certainly, the little finger would endeavor to overpower the hand, and the hand again would strive to overpower the little finger. The same may be said of two equal figures, as two hands, and other the like examples may be given. And this is also my meaning when I say that some shapes have power over others, and some degrees and temperaments of matter [have power] over others; whereby I understand nothing else but that some parts have power over others. Also, when I say that outward things govern, and a creature has no power over itself,[62] I mean that which is stronger, by what means soever, is superior in power. [. . .]

[532] [. . .] Neither would I have you to scruple at it, when I say that both parts or degrees of animate and inanimate matter do retain their own interior natures and proprieties in their commixture, as if those different natures and

---

60. A paraphrase of *Philosophical and Physical Opinions* (1663), part 1, chapters 13 and 14.

61. Cavendish paraphrases *Philosophical and Physical Opinions* (1663), part 6, chapter 3, where she writes that "infinite is a tyrant to motion, and motion to figure, and eternity to all."

62. Cavendish paraphrases *Philosophical and Physical Opinions* (1663), part 3, chapter 10. The actual passage is this: "But man, and for all that I know all other things, are governed by outward objects, they rule, and we obey; for we do not rule, and they obey, but everything is led like dogs in a string by a stronger power; but the outward power being invisible, makes us think we set the rules, and not the outward causes; so that we are governed by that which is without us, not that which is within us, for man has no power over himself."

proprieties, where one is self-moving and the other not, did cause them to be two different matters; for thus you might say as well that several figures which have several and different interior natures and proprieties are so many several matters. The truth is, if you desire to have the truest expression [533] of animate and inanimate matter, you cannot find it better than in the definition of nature, when I say nature is an infinite self-moving body; where by the "body" of nature I understand the inanimate matter, and by self-motion the animate, which is the life and soul of nature, not an immaterial life and soul, but a material, for both life, soul, and body are and make but one self-moving body or substance which is corporeal nature. [. . .]

[538] [. . .] Concerning the actions of nature, my meaning is that there is not any action whatsoever but was always in nature, and remains in nature so long as it pleases God that nature shall last, and of all her actions perception and self-love are her prime and chief actions; wherefore it is impossible but that all her particular creatures or parts must be knowing as well as self-moving, there being not one part or particle of nature that has not its share of animate or self-moving matter, and consequently of knowledge and self-love, each according to its own kind and nature; but by reason all the parts are of one matter and belong to one body, each is unalterable so far that, although it can change its figure, yet it cannot change or alter from being matter or a part of infinite nature; and this is the cause there cannot be a confusion amongst those parts of nature, but there must be a constant union and harmony between them, for cross and opposite actions make no confusion, but only a variety; and such actions which are different, cross, and opposite, not moving always after their regular and accustomed way, I name "irregular," for want of a better expression; but [539] properly there is no such thing as irregularity in nature, nor no weariness, rest, sleep, sickness, death, or destruction, no more than there is place, space, time, modes, accidents, and the like, anything besides body or matter.

When I speak of "unnatural motions,"[63] I mean such as are not proper to the nature of such or such a creature, as being opposite or destructive to it, that is, moving or acting towards its dissolution. Also when I call violence "supernatural," I mean that violence is beyond the particular nature of such a particular creature, that is, beyond its natural motions; but not supernatural, that is beyond infinite nature or natural matter.

---

63. *Philosophical and Physical Opinions* (1663), part 7, chapter 11.

When I say "a thing is forced,"[64] I do not mean that the forced body receives strength without matter; but that some corporeal motions join with other corporeal motions, and so double the strength by joining their parts, or are at least an occasion to make other parts more industrious. [. . .]

[542] [. . .] Thus, Madam, you have a true declaration of my sense and meaning, concerning those places which in my *Philosophical Opinions* you did note as being obscure; but I am resolved to bestow so much time and labor as to have all other places in that book rectified and cleared which seem not perspicuous, lest its obscurity may be the cause of its being neglected. And I pray God of his mercy to assist me with his grace, and grant that my works may find a favorable acceptance. In the meantime, I confess myself infinitely bound to your Ladyship, that you would be pleased to regard so much the honor of your friend, and be the chief occasion of it; for which I pray heaven may bless, prosper, and preserve you, and send me some means and ways to express myself, Madam,

Your thankful friend and humble servant.

---

64. *Philosophical and Physical Opinions* (1663), part 5, chapters 16 and 51.

Eternal God, infinite Deity,
Thy servant, nature, humbly prays to thee,
That thou wilt please to favor her, and give
Her parts, which are her creatures, leave to live,
That in their shapes and forms, whate'er they be,
And all their actions they may worship thee.
For 'tis not only man that does implore,
But all her parts, great God, do thee adore;
A finite worship cannot be to thee,
Thou art above all finites in degree.
Then let thy servant nature mediate
Between thy justice, mercy, and our state,
That thou may'st bless all parts, and ever be
Our gracious God to all eternity.

INDEX

accidents, xxx–xxxi, 37–40, 63

air: disease in, 146, 170; light and, 54–55; patterning in, 44, 50, 54, 67; perception of, 48, 61, 78

alchemy, xxxvii, xxxix. *See also* chemistry

angels, xxxvi, 109, 113–14, 205–6. *See also* supernatural beings

animals (nonhuman): no divine soul in, 32; generation of, 180–81; knowledge in, xxxiii, 71–72; language and, xxxii–xxxiii, 29, 71–72; memory in, 29; minds in, xxxii–xxxiii, 71; motions in, 35; perception in, 72; reasoning in, 32–34, 71–72; sense and reason in, 28, 108, 213–14

animal spirits, 70nn113–14, 118, 120–21

annihilation, 13, 38–40, 86, 192–93, 200

Antidote against Atheism, An (More), xxxiv, 81

antipathy. *See* sympathy/antipathy

appetites. *See* passions and appetites

Archelogia Philosophica Nova (Harvey, G.), xlv

archeus, xxxix–xl, 131, 133

Aristotle, xlii, 182n9, 194

art, 21–22, 29, 37, 79, 138–40. *See also* profitable arts

art of fire. *See* alchemy; chemistry

atheism, xxxv–xxxvi, 81–82, 86, 97, 123, 125, 152

atoms: Cavendish on, xliv–xlv; Charleton on, 196; as first principle, 13; Harvey (G.) on, 200–201; nonexistence of, 61, 74, 75, 196; in *Philosophical Opinions*, 219; in *Poems and Fancies*, xliv–xlv, 196–97

Boyle, Robert, x, xx–xxi, xxxix, xlvi, 202

Cartesian dualism. *See* dualism

causation: Cavendish on, xix–xx, xx n50, 53; mechanists on, xix; occasionalists on, xix, xxxi

Cavendish, Margaret Lucas: church, submission to, 17, 84, 127, 155, 157, 205; death, xv; early life, xi–xiii; education, lack of formal, xii, 4, 7, 40, 44, 174, 207; English Civil War and, xii–xiv; languages (non-English), inability to read, xxii, xxx–xxxi, xliii, 3–4, 203; marriage to Newcastle, xiii–xiv; in Netherlands, xiii–xiv; in Paris, xiii; reputation, vii–viii, 213; state, submission to, 17, 155; writings of, xii, xiv–xv

Cavendish, William. *See* Newcastle, 1st Duke of (William Cavendish)

Charleton, Walter, xliii–xliv, 194–96

chemistry, 113, 138–41, 150. *See also* alchemy

children, 33–34, 137, 156–57

Christianity: atheism and, 81, 152; Cavendish and, 16–17, 84, 127, 155; Cavendish on, 125–26, 152, 201, 204–5; divine soul and, 124; memory and understanding in heaven and, 159; origins of sin and death in 162–63

color, 14, 38, 45–46, 75–77

Conway, Anne, x n18, xxvii n106, xxxiv–xxxv

creation: generation distinct from, 216; God and, 13, 15–17, 53, 68, 84, 89, 199, 215, 218; nature and, 16, 38–40, 215–16

creatures, xvii–xix, 85–86, 175–76, 184; generation of, 179, 181, 216; God and, 81–82, 153; sizes of, 203–4; souls in, 5, 182–23. *See also* knowledge; motion; nature

deafness, 49, 107, 120

death, xix; animal spirits and, 118; as figure changes, 88, 157–58, 160, 173; as God's punishment for sin, 162–63; More on, 100; as motion changes, 44, 87, 123, 126, 170–71, 173; motions in dead things and, 100; same for men and women, 150; as separation of divine soul and body, 117–18; Van Helmont (J. B.) on, 162–63, 170

decay, 44, 118. *See also* death

deity. *See* God

Descartes, René, xxxi–xxxiv; on air, 74–75; on animals (nonhuman) and reason, xxxii–xxxiii, 71; on causation, xxxi; Cavendish reads in translation, xxxi, xxxi n80; on circular motion, xxxii; on elements, 68, 68n111; on matter, 68; as mechanist, xix–xxi, xxxiii–xxxiv; on mind, 70; on motion, xxxii, 62–64; Newcastle and xxxiii; on perception, xxxiii, 72–73; on place, 65; on soul, xxxv; on thunder, 75; on water, 74–75

devils, 109, 113–14, 205–6. *See also* supernatural beings

Dialogue Concerning the Two Chief World Systems (Galileo), xliii

digestion, 165–66

Discourse on Method (Descartes), xxxi, xxxi n80, 71

Discourses on the Virtuosi in France, The (anon.). *See General Collection of Discourses of the Virtuosi of France, A* (anon.)

disease: causes of, 169–70, 172–73; chemical treatment of, xl, 167–68; digestion and, 166; elements and, 135; God and, 162–63; immaterial spirits and, 119; as irregularity, xx, 160–62; as motion change, 87, 119–20, 146, 157–58, 160–62; plagues and, 119–20, 169–70; supernatural magic/witchcraft and, 146; Van Helmont (J. B.) on, 160–62; weakness and, 171–73

divinity, study of. *See* theology

dreams, xxxiv, 24–25, 25n39, 45, 47, 103

drugs, 168, 213

dualism, xxxii, 70

earth, 65–66, 117

echoes, 50, 53–54

eclipse, 117

elements, 13, 78, 108, 131, 135, 213–14

Elements of Philosophy, the First Section, Concerning Body (Hobbes), xxvii, xxix

Elisabeth, Princess of Bohemia, x, x n18, xxxvii, xxxvii n106

English Civil War: Cavendish's family in, xii–xiv

eternity: God and, 15–16, 61, 82, 218; infinity and, 11, 61; of matter, xvi–xxvii, 88, 200, 218; nature and, 16–17, 88, 218

experiments, 202–4

fairies, 203–4, 203nn45–46

fetal abnormalities, 168–69, 180

fetal development. *See* generation

figuring, xviii–xix, xxiii–xxiv, xxvi, 27–28, 45–46

fire, 45–46, 55, 78, 217

free will, xlvii, 61, 127, 162, 205–6

Galen, xxxviii, xli, 163

Galenism, xxxviii, 163, 167

Galileo, xli, xliii, 185–87, 189, 189n14

General Collection of Discourses on the Virtuosi of France, A (anon.), xlvi, 202–3

generation, xli–xliii; creation distinct from, 216; fetal abnormalities in, 180; divine soul and, 156–57, 182; Harvey (W.) on, xli–xliii, 178–81; in humans, 156–57, 181; life and, 131; metamorphosis and, 181; motions of matter and, 35, 39, 132, 158, 176, 178–83; "natural soul" and, 182–83; nature and, 91; Van Helmont (J. B.) on, 130–31; in vegetables, 181; women unsuited to study of, 175

God: as annihilator, 86; as creator, 13,
15–17, 53, 84, 89, 215, 218; as creator
of self-moving matter, 112–13, 115–16;
death and disease as punishment from,
162–63; eternity of, xxvi, 61, 82, 218; free
will and, 162, 205–6; Harvey (G.) on,
198–99; incomprehensibility of, 83–84,
153, 155, 218; as inconceivable by ideas,
82–83; infinity and immateriality of,
xxv–xxvi, 82–83, 151, 153, 198; More on,
83; nature and, xv–xvi, xix–xx,
xxv–xvi, 11–14, 37, 61, 96, 98, 198–99,
223; nature worships, 81–82, 153,
214–15; notion of, xxvi; as omnipotent,
92, 115–16, 151; in *Philosophical
Opinions*, 115, 220; purity and, 12;
religious conceptions of, 204–5, 218;
as supernatural, 11–13, 109, 218–19;
vacuum and, 194–95; Van Helmont
(J. B.) on, 151, 162
government, xxviii –xxix, 36–37

hand, 21, 39, 52–53, 73, 191–93, 220
Harvey, Gideon, xlv; on atoms, 200–201;
Cavendish on 197–201; on God,
198–99; life and work, xlv; on matter,
197–200; on nature, 199; theology
mixed with natural philosophy of, 201;
on universe as finite, xlv
Harvey, William, xli–xlii, 178–81
hearing. *See* sound
heat, 14, 45, 55, 165
Hobbes, Thomas, xxvii–xxxi; on accidents,
37–39; on appetites, 36; on body, xxix,
39; on color, 38; on free will, xlvii;
Newcastle and, xiii, xxviii; on dreams,
24–25, 25n39; on imagination, 23–26;
as mechanist, xix–xxi; on matter,
xix–xxi, 20; on memory, 23; on mind,
27; on motion(s), 20, 35–36, 60; on
perception, 17, 42–43; on place, 40–41;
on reasoning, 32–33; on scent, 58; on
sight, 49; on space, xxix, 60–61; on
speech and reasoning, 29; on sound, 49;
on understanding, 26, 30

humankind: atheism and, 81–82; divine
soul and, 28, 32, 128, 183; free will in, 61,
205–6; generation in, 181; God revealed
to, 153; knowledge in, 71–72; limitations
of, 32–33, 42, 87, 100, 149, 153, 213–14

iatrochemistry, xxxviii–xxxix. *See also*
medicine, chemical
imagination, xxiv, 23–26, 35, 102, 104, 212
immaterial spirits, 94; atheism and, 97; bodies
and, 14, 114, 118; Cavendish on, xxxvi;
disease and, 119; matter and, 104–5,
108–9, 112–13, 154–55; More on, xxxv,
xxxvi, 115; none in nature, 13–14, 88–89,
92, 104, 109–10, 154; rational motions
and, 105; as self-moving, 115; space and,
154; as supernatural beings, 154–55
inference: as argumentative strategy, xxi, xxxiv
infinities: nature and, 79–80, 216; types of,
9–12, 79–80
infinity: eternity and, 11; of God, xxv–xxvi,
82–83, 151, 153, 198; of matter, 64,
93–94; of motion, 64
Inquisition, the, xxxvii, xliii
irregularities, xx, xx n51, 82, 132, 187. *See
also* motion, irregular

knowledge: in animals (nonhuman), 71–72;
in creatures, 91, 96, 108, 122, 213–14;
as cause, 98; in humans, 71–72; natural
philosophy and, 208; as self-moving
matter, 112, 214; as sense and reasoning,
29; speech and, 31, 90; varieties of,
32–33. *See also* reasoning; sense and
reason; understanding

language, xxxii–xxxiii, 3–4, 26, 203, 218. *See
also* speech
letters, published, ix–xi, xi n18
Leviathan (Hobbes), xxvii–xxix, 17
light: air and, 48, 54–55; as a body, 14, 47;
color and, 45–46, 76–77; Descartes on,
72–73; figuring and, 45–46; of moon,
117; sight and, 45, 47, 56–57; of sun, 45,
47, 55. *See also* shadows

lodestone, xl, 142, 144, 213. *See also* magnetism

madness, xxiv, 50, 91, 146
magic, xxxvii, xl, 145–46
magnetism, xl, 141, 146
mankind. *See* humankind
matter: animate and inanimate mixed, 63, 70, 86, 211–12, 220–21; annihilation and, 13, 86; antipathy/sympathy and, 192; as corpuscular, xxxiii–xxxiv; degrees of, xvii, 69, 114, 130, 210–11; divisibility of, 104–6; eternity of, xxvi–xxvii, 84, 88, 115, 200; figure and motion inseparable from, 13–14, 63, 88, 93–93, 212; generation and, 176, 178–81; Harvey (G.) on, 198–200; Hobbes on, 20; infinity of, 93–94, 188, 198; mechanists on, xix–xxi, xxxi; motion inseparable from, xvii, 179, 190–92; More on, xxxv, 89–90, 93, 100; occasionalists on, xxi; in *Philosophical Opinions*, 211, 220–21; perfection and, 188; purity of, 69, 176–77; as self-moving, xx–xxi, 20, 61, 64, 69, 77, 88–92, 115–16, 175–76; supernatural not mixed with, 13, 86; weight and, 217
matter, animate, xvii; as eternal, 217; perception and, 100–101; purity and subtlety in, 129–30; as self-moving, 64; sense and reason in, 29–30, 213–14
matter, inanimate, xvii; animate matter and, 190–91, 200, 211; Harvey (G.) on, 200; movement and, 63; in nature, 79; purity of, 177; rest and, 210–11; sensitive matter and, 101, 212
matter, rational, xvii; appetites and, 105, 110; different quantities in different bodies, 111, 130, 185; divisibility of, 106; ideas as, 210; imagination as rote movement of, xxiv, 102, 104, 212; immaterial spirits and, 105; memory and, xxiv, 102, 106, 212; mind as, 102; "natural" soul as, xxxii, 70; passions and, 101, 103, 212; purity and subtlety

of, 73, 101, 104, 106, 129–30, 176–77, 193, 210; reason and, 72; sensitive matter, differences from, 70, 89, 101–3, 107; understanding and, 106, 212
matter, self-moving: echoes as, 53–54; God as creator of, 112–13, 115–16; knowledge and, 112, 214; magic and, 145–46; More on, xxxvi, 89–90, 95, 97, 99; perception and, xxii–xxiv, 20–22, 112; reflections as, 54; sense and reason as, 31, 89–91, 148–49; understanding and, 26
matter, sensitive, xvii; density and rarity in, 210; dreams as rote movement of, xxiv, 24–25; inanimate matter and, 101, 212; life and, 210; madness as rote movement of, xxiv, 50; purity and subtlety in, 130, 176–77; rational matter, differences from, 71, 101; senses and, 91
mechanism, xix–xxiii, xxxi, xli
medicine, xxxviii–xxxix, 163–65, 167–68
medicine, chemical, 140, 168. *See also* iatrochemistry
medicine, Galenic. *See* Galenism
memory: in heaven, 159; Hobbes on, 23; rational matter and, xxiv, 27, 102, 106, 111, 212; speech and, 29
mind: body inseparable from, 70; differences in, 96; divisibility of, 85; Hobbes on, 27; materiality of, 108–9, 111; rational matter as, 102. *See also* dualism; matter, rational
minimas. *See* atoms
More, Henry, xvi, xxxiv–xxxvii; on animal spirits, 120–21; on atheism, 81; Cavendish on, xxxv–xxxvi, 120–23, 125; Conway and, xxxiv–xxxv; on death and decay, 100, 117–18; Descartes and, xxiv; on divine soul, 114–15, 117–19, 120–21; on God, 83; on immaterial spirits, 114; on matter, xxxv–xxxvi, 89–90, 93, 95, 97, 99–100; on motion, 95, 100, 120–21; on mind, 108–9; on nature's laws, 87; on passions, 119; on perception, 102, 104;

reputation of, xxxiv; on soul, xxxv–xxxvi; writings of, xxxiv

motion, xviii–xx; animal, 35; circular, xxxii, xliii, 64, 68, 186–87; creatures and, xviii, 148–50; in dead things, 100; degrees, not by, 189; Descartes on, xxxi–xxxii, xxxv, 62–64; divine soul and, 114–15, 120–21; external, xviii, 73–74, 190, 216–17; forced, 21–24, 149–50, 165–66, 222; Galileo on, 185–87, 186n11, 189; generation and, xli, 179–81; Hobbes on, xxxv, 20; internal, xviii, 216–17; matter inseparable from, 14, 52, 62–63, 73–74, 179, 190–92, 217; More on, xxxv, 95, 100; occasioned, 149–50; patterning and, 101, 212; in *Philosophical Opinions*, 222; power of, 222; regular, xx; rest as change in, 185–86; senses and, 98–99, 171, 176–77, 193; sensitive, 47–50, 59, 78, 101, 120, 130; sympathy/antipathy as, 102; time and, 147, 185, 196, 212; not transferable without matter, 52–53, 55, 60, 62–64, 190–93; varieties of, 64–65, 148–50; vital, 35; weight and, 217

motion, irregular, xx; birthmarks and, 137; blindness and, 107; deafness and, 49, 107; disease as, 160–62; falsities and, 31; fetal abnormalities and, 169; madness and, 50; muteness and, 107

natural philosophy: Cavendish on, 8, 175; chemistry and, 138–41; knowledge and, 208; probability and, 207–8; profitable arts and, 208; theology distinct from, xxvii, 8, 122–26, 150–53, 155, 201; of Van Helmont (J. B.), 130–31, 133

nature: annihilation and, 28, 38–39, 86; Cavendish on, xv–xvi; as creator, 16, 38–40, 132, 215–16; as creatures, collection of, xvii–xviii, 181, 209; contraries in, 136–37; as eternal, 16–17, 88, 200, 218; figures and forms in, 69, 85, 88, 158, 200; free will in, 127, 206;

generation and, 91; God and, xv–xvi, xxxv, 11–14, 37, 61, 86, 96, 198–99, 223; God worshipped by, 81–82, 86, 214–15; Harvey (G.) on, 199; immaterial spirits and, 88–89, 97; infinity of, 4, 11–12, 79–80, 82–83, 198–99, 216; irregularities in, 98, 187; laws of, 87; materiality of, 5, 11–12, 14, 69, 82–83, 85–86, 134, 184; matter and, xv–xvi, 77–78, 132, 188; mistakes in, 209; motions in, 14, 68–69, 96; perception in, xxiv–xxv, 209; perfection in, 188; in *Philosophical Opinions*, 219, 221; rest and, 41–42, 64, 97–98; soul and, xvi, 184; not supernatural, 126, 194; sympathy/antipathy in, 93, 95; Van Helmont (J. B.) on, 133, 137; variety in, 28, 85, 116, 148–49, 161; as wise, 69, 85–87, 90–92, 95–97, 176, 208

Newcastle, 1st Duke of (William Cavendish): Descartes and, xxxiii; English Civil War and, xiii–xiv; Hobbes and, xiii; Hobbes-Bramhall debate on free will and, xlvii; marriage to Cavendish, xiii; natural philosophy and, xiii

notions, xxvi

Of the Immortality of the Soul (More), xxxiv, 110

Paracelsus, xxxviii–xxxix, 163

passions and appetites, 36, 103, 105, 110, 119, 141–44, 190, 212

patterning, xxiii–xxiv; in air, 67; figures and, 51; hearing as, 49–50; hand and, 52, 73; motion and, 101, 212; perception as, 29, 44; reflections and, 54, 56–58; scent and, 58–60, 78; in shadows, 116–17; sight as, 20–21, 78; in snow, 67; sound and, 49–50, 54, 78; of sunlight by air, 55; taste and, 78; no translation of matter in, 179, 182. *See also* figuring

Pepys, Samuel: on Cavendish, vii

perception: in animals (nonhuman), 72; animate matter and, 100–1; Descartes on, xxxiii, 72–73; double, xxiii–xxiv, xxxiii, 18, 50, 73; in elements, 214; exterior motion and, 73–74; Hobbes on, xxi–xxii; More on, 102, 104; nonhuman, xxiv–xxv; objects not subject to, 98–99; patterning and, 29–30, 44, 51, 78; pressure, not by, xxii, 18–19, 20–22, 26, 30, 42–44, 73, 101–3; rational, xxiii–xxiv; rational and sensitive motions and, 18, 73–74, 78; reaction and, 102–3; reason and, 72; self-motion as, 43–44; self-moving matter and, xxii–xxiv, 20–22, 91, 112; sense organs and, 214; sensitive, xxiii–xxiv, 101; sight as, 18–19, 209; sound and, 19; in vegetables, 214

philosopher's stone, xxxix, 139–40, 208

Philosophical and Physical Opinions (Cavendish). See *Philosophical Opinions* (Cavendish)

Philosophical Letters (Cavendish): Cavendish on, xv, 3–4, 8, 129, 174, 207; as genre, ix–x; *Philosophical Opinions* and, xv, 4–5, 206–7; seventeenth-century philosophy and, ix, xlvii; significance of, ix–xi, xv

Philosophical Opinions (Cavendish), xiv; on atoms, xliv–xlv, 219; Cavendish on, 7–8, 174, 206–7, 219–22; on color, 77; critical reactions to, 7–8; on disease, 120; on dreams, 25, 25n40; fairies in, 203, 203n45; on forced motion, 222; on God, 115, 220; on knowledge, 32; on matter, 64, 115, 211, 220–21; on motion, 64, 220–21; on nature, 188, 219–21; on passions, 119; on perception, 18–19; *Philosophical Letters* and, xv, 4–5, 206–7; on vacuum, 215, 215n53

place: Cavendish on, xxx; Descartes on, 65; body inseparable from, 40–42, 47, 60, 65–68, 94, 154, 189, 212, 215; Hobbes on, xxx, 40–41; immaterial spirits lack, 154; magnitude and, 41

Poems and Fancies (Cavendish), xiv; atoms in, xliv–xlv, 196–97; fairies in, 203–4, 203n46

political theory. See government

predestination. See free will

Princess Elisabeth of Bohemia. See Elisabeth, Princess of Bohemia

profitable arts, 202–4, 208

purity: in animate matter, 129–30; degrees of, 79; generation and, 180; God and, 12; of matter, 69–70, 99, 176–77; in rational matter, 73, 101, 104, 106, 129–30, 210; in sensitive matter, 130, 176–77

rarefaction, 74–75

reasoning, 29–34, 71–72, 158–59

reflections, 50, 54, 56–58

religion, 123–25, 204–5, 218. See also atheism; Christianity

remembering. See memory

reproduction, process of. See generation

rest: Galileo on, 198n14; inanimate matter and, 210–11; motion and, 185–86; nature, none in, 41–42, 64, 86, 97–98, 193

Royal Society of London: Cavendish visits, vii, xlvi

scent, 14, 58–60, 78

scholastic philosophy, xxix–xxx, 197

self-love, 28, 221

sense and reason: in animals (nonhuman), 28, 108, 213–14; different quantities in different creatures, 130; in drugs, 213; in elements, 108, 213; in lodestones, 213; as self-moving matter, 31, 89–91, 148–49; speech and, 34; in vegetables, 34, 108, 213

sensory perception. See perception

shadows, 116–17

sight, 18–21, 45–48, 56–58, 76–78, 120, 209

sleep, 21, 45, 87. See also dreams

smell. See scent

snow, 67, 74–75

soul, xvi, xxxii, xxxv–xxxvi, 129

soul, divine, xvi, xxxii, 70, 85, 96; animal spirits and, 120–21; body and, 109–10, 128; generation and, 156–57, 182; in humans, 28, 32, 128, 183; immortality

of, 117–18, 122–24, 126, 167; More on
114–15, 117–19, 120–21; indivisibility
of, 114–15, 157; motion of matter
and, 114–15, 120–21; "natural" soul,
differences from, xvi, xxxii, 70, 85, 106,
122–23, 126; pain and, 157; passions
and, 119; production of, 156–57; reason
and, 159; vacuum and, 194–95; Van
Helmont (J. B.) on, 156
soul, "natural": as animate matter, 129–30;
divine soul, differences from, xvi, xxxii,
70, 85, 106, 122–23, 126; divisibility of,
184; generation and, 182–83; immortality
and, 126; materiality of, 14, 70, 106, 126;
reason and, 159. *See also* mind
sound, 14, 19, 49–51, 54, 75, 78
space, xxix–xxx, 41, 60, 154, 189, 196
speech, 29–31, 34, 71–72, 90. *See also*
language
supernatural beings, 17, 52–53, 109,
127–28, 154–55. *See also* angels; devils;
soul, divine
sympathy/antipathy, 26, 93, 95, 102, 141–44

theology: Cavendish on, 3, 8; natural
philosophy distinct from, xxvii, 8,
122, 124–26, 150–53, 155, 201; Van
Helmont (J. B.) and, xxvii
thunder, 75
time, 94–95, 147, 185, 196, 212

understanding, 26, 30–31, 85, 106, 159,
212. *See also* knowledge; mind

vacuum, xliv–xlvi, 194–95, 215, 215n53
Van Helmont, Franciscus Mercurius, xxxviii

Van Helmont, Jan Baptist, xxxvii–xli;
as chemist, 138–39; Cavendish on,
xl–xli, 129, 131–34, 136–37; on death,
162–63, 170; on disease, xl, 160–62;
education and life, xxxvii; on elements,
135; on Galen and Galenism, 163, 167;
on generation, 130–31; on God, 151,
162; iatrochemistry and, xxxviii, 163;
natural philosophy of, 130–31, 133; on
nature, 133, 137; on Paracelsus, 163; on
plague, 169; on reason and reasoning,
158; on soul, 129, 155; theology mixed
with natural philosophy by, xxvii, 136,
150–51; on time, 147; on water, xxxix–
xl, 130–31, 136; weapon-salve debate
and, xxxvii; on women, xli, 134, 136–37,
150, 168–69

vegetables, 34, 108, 169, 181, 213–14

water, xxxix–xl, 62, 65–66, 74–75, 78,
130–31, 136
weapon-salve, xxxvii, xl, 142
witchcraft, 145–46. *See also* magic
witches, 134, 145
women: birthmarks in children and,
136–37; death in, 150; education, lack of
formal, 207; generation an unfit subject
for study by, 175; government an unfit
subject for study by, xxviii, 36; letters,
published and, x–xi, x n18; supernatural,
targets for fear of, xl–xli; Van Helmont
(J. B.) on, xli, 134, 136–37, 150, 168–69;
as witches, 134, 145
Woolf, Virginia: on Cavendish, viii
Worlds Olio, The (Cavendish): critical
responses to, 5–6